D0113810

ELEMENTARY LOGIC

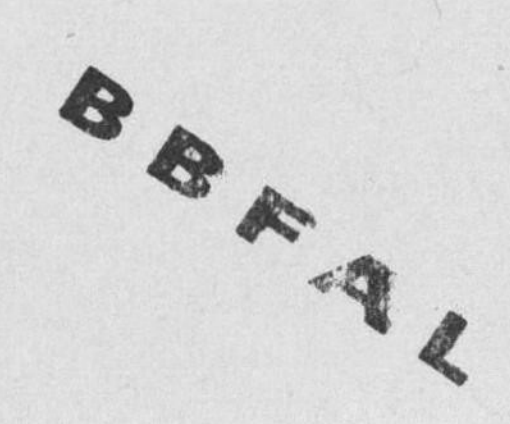

ELEMENTARY LOGIC

Second Edition

BENSON MATES

DEPARTMENT OF PHILOSOPHY
UNIVERSITY OF CALIFORNIA

New York OXFORD UNIVERSITY PRESS 1972

Fourth printing, 1977

LIBRARY OF CONGRESS CATALOGUE CARD NUMBER: 74-166004
PRINTED IN THE UNITED STATES OF AMERICA

PREFACE TO THE FIRST EDITION

Elementary logic, in the technical sense of the term, is that part of symbolic or mathematical logic in which the notions of 'all' and 'some' are applied only to individuals and not also to classes or attributes of individuals. It is at once the most solid part of the subject and the simplest part that permits a wide variety of non-trivial applications. Consequently, it seems especially suitable for study by beginners.

The present textbook is intended as an introduction to elementary logic. Its content, structure, and manner have been determined in large measure —perhaps 'caused' is the better word—by certain desiderata about which the reader should be informed at the outset. The leading idea is that even an introductory treatment of logic may profitably be fashioned around a rigorous framework. Students come to this subject with the legitimate expectation that here, if anywhere, clarity and exactitude will prevail. When they meet sloppiness or vagueness at crucial places, they understandably feel let down. While of course I do not imagine that I am the first textbook writer to cherish rigor, or still less that I have succeeded in giving no occasion for the sort of disappointment just mentioned, my efforts in that direction will account for many of the book's distinctive features.

As a first step toward achieving the desired level of clarity, I have endeavored in the formal developments to supply a reasonably exact definition for each of the principal concepts involved, including those of *sentence, interpretation, truth* (with respect to an interpretation), *consequence, validity, consistency, tautology,* and *derivation.* The presence of these definitions has set the style of the book. Typically the discussion of a given topic consists of a brief introduction, followed by definitions, followed by numerous examples and explanatory comments. At first reading, no doubt, some of the definitions will be found rather opaque, but it is hoped that the supplementary comments and examples will render them intelligible. At any rate, they are available for reference and will at least enable the instructor to answer questions about what is meant.

This book does not pretend to be sufficient for the autodidact. It is designed only to present a core of basic material that is suitable for use by a competent instructor in any of a wide range of possible introductory courses. Additional explanation of the fundamental concepts will surely be necessary; further examples of formalized theories should be presented, the choice depending upon the interests and background of the students; and no introductory treatment of the subject should fail to include a general discussion of the logistic method. Every instructor will think of further items to add; if at the same time he does not find too much that has to be subtracted, the book may serve its purpose.

The first two chapters are devoted to preliminaries needed for the formal developments that follow, and to certain philosophical points that might be described as constituting an 'intuitive background' for those developments. Chapters 3 and 4 describe the grammar and semantics of a rather simple artificial language and lead to a definition of *logical consequence* for sentences of that language. The associated definition of *truth* for these sentences makes it possible to avoid the fundamental confusion generated when one 'defines' the so-called sentential connectives with the help of truth tables and then uses these connectives between open formulas (as in '$(x)(Fx \vee Gx)$'), while at the same time admonishing the student that only sentences (closed formulas) can be true or false.

The fifth chapter, which is perhaps the least satisfactory in the book, attempts to throw some light on the problems involved in 'symbolizing' English sentences by means of the artificial language. Here special care is necessary if one is to avoid saying things that are misleading or downright false. I have felt obliged to reject the approach in which connectives and quantifiers of the formal system are regarded as abbreviations for corresponding words and phrases of the natural language, and in which the formal sentences are described as schemata for sentences of the natural language. Characteristic of this approach is the suggestion that logical words like 'if . . . then', 'or', and 'not' are vague and ambiguous when used in everyday speech, and that what the logician does is to select in each case a single sense and bestow that sense upon the formal counterpart. Such claims are poorly substantiated, and, in my opinion, probably false. In general I have tried to make it quite clear that although to some extent we can translate natural-language sentences into our formal notation there is frequently a serious amount of slippage in the process. Not only is it possible for a translated argument to be formally unsound while the original is intuitively sound, but also there are cases in which the translated version is formally sound while the original is intuitively unsound. The price of drawing explicit attention to this fact is, I suppose, loss of the gratifying and perhaps pedagogically useful impression that logicians possess an esoteric apparatus for testing the soundness of arguments framed in the natural language.

Chapter 6 deals with formal tautologies. It seeks to provide the student with a clear procedure for deciding whether an arbitrary sentence of predicate logic is tautologous, and it introduces a very simple deductive system that is intended to acquaint him with large numbers of tautological sentences as well as to give him practice in deduction.

Chapter 7 presents a similar but more complex natural-deduction system for the entire predicate calculus. In designing this system I had in mind the following considerations, among others:

(a) Logic seems most naturally conceived as a system of inference rules that are typically applied to extralogical subject matter, rather than, for example, as an axiomatic system of logical truths. It is easy to verify that in actual argumentation, on any subject from politics to mathematics, the use of logic seldom involves the actual insertion of logical truths into the argument. For this and other reasons I have decided to place primary emphasis upon natural-deduction systems, as contrasted with the axiomatic approach.

(b) An argument, in the fundamental sense of that term, is a system of declarative sentences, not a system of conditions or open formulas. Only sentences can have truth values, and only sentences can, in the first instance, stand to one another in the relation of logical consequence. This, together with certain less important considerations of technical convenience, accounts for the decision not to allow formulas with free variables to appear on the lines of a derivation.

(c) Since traditionally the chief purpose of any system of logic is to provide a way of reducing complex inferences to chains of simpler ones, it seems desirable that each basic inference rule should be sound, in the sense that when applied to true premises it yields only true conclusions. This desideratum excludes various convenient rules that might otherwise have been introduced.

(d) Closely related to the foregoing, a sound argument should be so constructed that if it is interrupted at any point the last sentence asserted follows logically from the premises utilized in obtaining it. In other words, every initial portion of a sound argument should also be a sound argument. This demand is automatically satisfied if (c) is satisfied, and in practice its satisfaction gives rise to numerous advantages. For instance, by noting at each step whether what he has just written is intuitively a consequence of its premises, the beginner who is learning such a system can always keep a rough check on whether he has applied the inference rules correctly. Also, the soundness of this type of system is most easily seen; since no single rule can lead from true premises to a false conclusion, it is obvious that the same will hold when the rules are applied consecutively.

The eighth chapter sets forth some of the general properties of the inference rules given in Chapter 7. It may be omitted without loss of continuity, though the material on duality and normal form will give the student valuable practice in the manipulation of quantifiers. In Chapter 9 the artificial language, which up to that point has been kept as simple as possible in order to focus attention upon the essentials, is expanded to include the apparatus of identity and operation symbols. Chapter 10 is devoted to an axiomatic presentation of elementary logic (with identity). Its inclusion reflects the historic interest of logicians in the fact that the principles of logic may themselves be arranged into a deductive system. In Chapter 11 examples of formalized theories are given. I have included a version of Aristotle's theory of the syllogism, following as closely as possible his own presentation of the subject. The last chapter is historical; it contains, among other things, a literal translation of the Aristotelian text upon which the formalization in Chapter 11 is based. I hope that this historical material will help to shield the student from the mistaken notion that modern symbolic logic is a newfangled fad somehow usurping the rightful position of what is called 'traditional logic'. The latter, curiously enough, has a 'tradition' that is relatively short, while many of the distinctive methods and ideas of modern logic (e.g., formalization and the use of variables, axiomatization of logical truths, construction of complete systems of inference rules, and so forth) are traceable back to the times of Aristotle and the Stoics, twenty-three centuries ago.

I wish to express my gratitude for the encouragement and assistance that have so generously been given me by my friends and colleagues, especially by David Rynin, Ernest Adams, William Craig, John Myhill, David Shwayder, Leon Henkin, Richard Montague, Donald Kalish, Rudolf Grewe, and Leon Miller, and most especially by Gebhard Fuhrken and Dana Scott. The very considerable extent to which I have drawn from other textbooks, particularly those of W. V. Quine, Alonzo Church, Patrick Suppes, and Irving Copi, will be evident to anyone familiar with these works. In the historical chapter I have leaned heavily upon the writings of William and Martha Kneale, Philotheus Boehner, Ernest A. Moody, Alonzo Church, and I. M. Bocheński. My translation of Aristotle on pages 208–9 is quoted by permission of the Encyclopædia Britannica.

Most of all, I wish to record my indebtedness to Alfred Tarski. The depth, breadth, and clarity of Tarski's work in logic are a source of inspiration to every serious student of the subject. Ideas of his, in however modified or attenuated a form, will be found throughout the book.

BENSON MATES

Berkeley, California
October 1964

PREFACE TO THE SECOND EDITION

In this revision I have tried to correct a number of errors and infelicities that became evident to me after publication of the first edition. I hope that the corrections have not introduced new difficulties, as can all too easily happen.

One of the more serious confusions in the first edition concerned the notions of 'direct occurrence' and 'sentence-form', and was pointed out to me by my colleague, Professor George Myro. If we ask whether 'Necessarily, $x < 7$' is a sentence-form, we find that it is obtainable from 'Necessarily, $8 < 7$' in which (on my criterion) '8' occurs directly, as well as from 'Necessarily, $4 < 7$', in which '4' does not. Thus, 'Necessarily, $x < 7$' would turn out to be a sentence-form, although it is a paradigm of the kind of formula we wish to exclude from use in specifying sets. The requisite emendation here, that a sentence-form be obtainable from a sentence by replacing direct occurrences, etc., and *not be obtainable from any other sentence by replacing indirect occurrences,* etc., would remove whatever value the concept had for enabling the student in practice to sort out the 'bad actors' among possible sentence-forms; and I have concluded that this purpose is better served by simply following the Fregean approach, in which names are said to occur indirectly if they are within the scope of such expressions as 'knows that', 'believes that', 'wonders whether', *and the like.*

Most of the other errors that have been rectified are of a technical nature. In response to many requests, the number and variety of exercises have been considerably increased. Also, a section has been added in which a proof procedure is set forth for the first order predicate calculus.

Readers familiar with the first edition may be puzzled as to why I would think it worthwhile to go through the entire book replacing the term 'analytic sentence' by the term 'necessary truth', when the two seem completely synonymous. My reason is that, in view of the history of 'analytic' as a technical term in philosophy, its recent use as a straight synonym for 'necessarily true' is inept and unfortunate. According to Kant, whose thought in this connection was obviously influenced by Leibniz, a judgment '*A* is *B*' is analytic if the predicate *B* is contained ('perhaps covertly', as he says) in the concept *A*; otherwise it is synthetic. In other words, his idea is that '*A* is *B*' is analytic if 'analysis of the concept of the subject, *A*, reveals the predicate, *B*'. Use-mention and other confusions aside, it is clear that this sense of 'analytic' does not apply to necessary truths like 'Either Aristotle taught Alexander or Aristotle did not teach Alexander', and still less to cases like 'If all men are mortal and all Greeks are men, then all Greeks are mortal'. The need for preserving the distinction between analyticity and necessary truth becomes even more apparent if one considers the view of Leibniz that some analytic propositions are false. According to him, '*non entis nullae proprietates sunt*'—'what does not exist has no properties'—so that '*A* is *B*' is false if *A* does not denote, whether or not the concept expressed by *B* is contained in the concept expressed by *A*. Such a view is not unheard of today; that e.g. 'The greatest positive integer is greater than any other positive integer' is false although 'its predicate is contained in its subject' is at least an arguable philosophical position, which may be obfuscated if we persist in using 'analytic' to mean the same as 'necessarily true'.

In remarking that propositions, statements, thoughts, and judgments, as described by those who consider them to be what logic is all about, do not exist, I seem to have stamped on the toes of a number of readers. This remark, some have maintained, has no place in an introductory logic text, is not consonant with the generally restrained tone of the book, is not accompanied by supporting argumentation, and irritates thoughtful students and instructors alike. In fact, I suppose that almost the only thing that can be said for it is that it is *true*. At any rate, let me note that it occurs in a section in which my purpose is to acquaint the reader with the point of view from which the book is written, and in relation to that goal I have no doubt that its inclusion is justified.

The list of scholars who have kindly offered me their criticism and suggestions is now even longer than it was before, and I am most grateful to all of them. Particularly helpful have been the comments of William Craig, Dagfinn Føllesdal, Veronica Karp, David Keyt, Donald Monk, George Myro, and Ian Mueller, to whom I extend correspondingly emphatic thanks.

BENSON MATES

Berkeley, California
May 1971

CONTENTS

ELEMENTARY LOGIC

1

INTRODUCTION

1. *What logic is about*
2. *Soundness and truth*
3. *Soundness and necessary truth*
4. *Parenthetical remarks*
5. *Logical form*
6. *Artificial languages*

This chapter is designed to give an informal and intuitive account of the matters with which logic is primarily concerned. Some such introduction is surely required; otherwise, the beginner is likely to feel that he does not get the point of the formal developments later introduced. Yet at the same time it is necessary to acknowledge the fact that logicians do not agree among themselves on how to answer the seemingly fundamental questions here treated. As concerns the formal developments there is remarkably close agreement, but any question bearing on 'what it's all about' tends to bring forth accounts that are very diverse.

For instance: Is logic about the way people think, the way they ought to think, or neither of these? Is it principally concerned with language or with the extralinguistic world? Are the logician's artificial languages to be regarded as simplified but essentially faithful *models* of natural languages, or are they to be thought of as proposed *replacements* for natural languages, or is their utility to be explained in some other way?

No doubt issues like these must eventually be faced, despite their vagueness. The beginning student should realize, however, that in practice their importance is not as great as one might suppose. When the neophyte asks 'What is mathematics?' or 'What is physics?' perhaps the best reply is: 'You can make up your own mind as to that, *after* you have become acquainted with what mathematicians and physicists do'. The same general point holds true of logic. Thus although we hope that our informal discussion will smooth the way for understanding the later technicalities, we are also aware that after the latter have been mastered the student may, guided by his own philosophical lights, decide to reject as false, circular, or even unintelligible certain portions of the introductory account now to be given.

1. *What logic is about.* Logic investigates the relation of *consequence* that holds between the premises and the conclusion of a sound argument. An argument is said to be *sound* (correct, valid) if its conclusion *follows from* or is a *consequence of* its premises; otherwise it is *unsound.* In some cases, notably the paradigms that appear in traditional logic books, the soundness or lack of soundness is immediately obvious. Nobody has much difficulty in seeing that the conclusion of the argument

All men are mortal;
All Greeks are men;
Therefore, all Greeks are mortal

follows from its premises. In other cases a little thought is required, as in

There are exactly 136 boxes of oranges in the warehouse;
Each box contains at least 140 oranges;
No box contains more than 166 oranges;
Therefore, there are in the warehouse at least six boxes containing the same number of oranges.

And in still other cases the question can be very difficult indeed. For instance, nobody has yet been able to make the discovery needed for deciding whether the one-premised argument

The number of the stars is even and greater than four;
Therefore, the number of the stars is the sum of two primes

is valid.

Note that in the last of these examples, which in essential respects is typical of a very large class of cases, the difficulty does not arise from such vagueness as may exist in the premise or in the conclusion. The crucial statement that every even integer greater than four is the sum of two primes seems to be *understood* well enough; our only problem is to discover whether it is *true.* Of course we frequently do meet arguments the soundness of which cannot be assessed because of vagueness or ambiguity. The clarification of meaning is sometimes a useful intellectual activity, and indirectly logic can be of considerable help in that connection, but nevertheless it is worth emphasizing that the question whether a given argument is sound is not automatically answered as soon as one has a clear idea of what the premises and conclusion mean.

2. *Soundness and truth.* By an *argument* we mean a system of declarative sentences (of a single language), one of which is designated as the *conclusion* and the others as *premises.* Thus the first example given above is a system of three declarative sentences, with the last designated as conclusion and the other two as premises; the second consists of four sentences, again with the last as conclusion; and in the third case we have an argument consisting of a single premise and a conclusion. Actually there is no reason not to consider cases in which the number of premises is infinite, though perhaps in that event the term 'argument' is no longer appropriate. For instance, we may ask whether or not the sentence

Every positive integer is less than the number of the stars

follows from the (infinitely many) sentences

1 is less than the number of the stars,
2 is less than the number of the stars,
3 is less than the number of the stars,
.
.
.

i.e., from the class of all sentences obtainable from the expression '*x* is less than the number of the stars' by replacing '*x*' by an Arabic numeral denoting a positive integer. As we shall see later, it is also useful to consider cases in which the number of premises is zero.

A *sentence* is defined by traditional grammarians as a linguistic expression that states a complete thought. Despite the obvious drawbacks of this definition, it is adequate for our present purpose, which, be it remembered, is only to give the student a preliminary intuitive acquaintance with the subject matter of logic. Sentences are usually classified as declarative, interrogative, imperative, etc. Characteristic of declarative sentences is that they are true or false, and hence it is these that are of primary interest to the logician.

The criterion for the soundness of an argument is commonly formulated in terms of the concepts of truth and possibility—about both of which we shall have more to say later—in the following way: an argument is sound if and only if it is not possible for its premises to be true and its conclusion false. Of course the term 'possible' is crucial here. We are not concerned with whether the premises or conclusion *are* in fact true; all that is required for soundness is that *if* the premises were true, the conclusion would have to be true. Another way of stating the same criterion is this: an argument is sound if and only if every conceivable circumstance that would make the premises true would also make the conclusion true.

Thus, although of course there are sound arguments in which the premises and conclusion are true, as in the first example given in section 1 above, there are also perfectly sound arguments having one or more false premises and/or a false conclusion. For instance, all component sentences of the sound argument

All men are clever;
All primates are men;
Therefore, all primates are clever

are false. Nor do sound arguments always lead from false premises to false conclusions; e.g.,

All Senators are old;
All octogenarians are Senators;
Therefore, all octogenarians are old

is sound even though it has false premises and a true conclusion. In each of these cases, regardless of what the truth-values of the components may be in fact, we see that if the premises *were* true the conclusion would have to be true, and this is enough to ensure soundness. The only combination of truth-values that *cannot* occur in a sound argument is that the premises be true and the conclusion false. For if the premises of an argument are in fact true and the conclusion false, it is (to say the least) possible that the premises should be true and the conclusion false, and hence the argument is unsound.

Unsoundness, on the other hand, can occur with all combinations of truth-values. Here are examples to illustrate some of the possibilities:

Some Senators are old;
Some generals are Senators;
Therefore, some generals are old

(This is an unsound argument with true premises and true conclusion);

Some professors are Swedes;
Some Norwegians are professors;
Therefore, some Norwegians are Swedes

(Unsound argument with true premises and false conclusion);

Whoever understands the subject gets a high grade;
Therefore, whoever gets a high grade understands the subject

(Unsound argument with false premise and false conclusion);

Whoever lives in Europe's oldest democracy lives in Zurich;
Therefore, whoever lives in Zurich lives in Europe's oldest democracy

(Unsound argument with false premise and true conclusion).

All of this shows that soundness in an argument does not depend simply upon the truth-values of premises and conclusion. It guarantees only that *if* the premises are true, then the conclusion is also true; it does not guarantee that any of the premises are in fact true, nor does it give us any information about the truth-value of the conclusion in case one or more of the premises is false.

3. *Soundness and necessary truth.* Closely connected with the concept of a sound argument is that of a *necessary truth.* A sentence (and from this point onward let us use 'sentence' as short for 'declarative sentence in written form') is a necessary truth if and only if there are no conceivable circumstances of which it would be false. The sentence

Socrates died in 399 B.C. or Socrates did not die in 399 B.C.

is a necessary truth, for it is true independently of the facts about Socrates or about anything else. By contrast, the sentence

Socrates died in 399 B.C.,

while true, is not necessary. There are easily describable circumstances of which it would have been false.

We shall return in a moment to the notion of a necessarily true sentence, but first let us examine the connection between necessary truth and soundness. To state this relationship compactly a bit of special terminology is helpful. A sentence of the form

If . . . then - - -,

where the blanks are to be filled with other sentences, is called a *conditional;* the sentence that immediately follows the word 'if' is called the *antecedent,* and what follows the word 'then' is the *consequent.* Now, given an argument with finitely many premises, we construct the so-called *corresponding conditional* by taking as antecedent the sentence that results from conjoining all the premises by means of the word 'and', and as consequent the conclusion. Then the connection between soundness and necessary truth is just this: an argument with finitely many premises is sound if and only if the corresponding conditional is necessarily true.

For example, the first argument given in section 1 is sound if and only if the conditional

If all men are mortal and all Greeks are men, then all Greeks are mortal

is a necessary truth, which it is.

Instead of referring to '*the* connection' between soundness and necessity, we should have said '*a* connection', for another interesting relationship is the following: a sentence is a necessary truth if and only if there is no unsound argument of which it is the conclusion. In other words, a necessarily true sentence is a consequence of *every* set of premises, and any such sentence is a necessary truth. For suppose that a sentence *S* is necessarily true, and consider an argument having *S* as conclusion. Now the argument is sound if and only if it is not possible that its premises be true and its conclusion, which is *S*, false. But since *S* is necessarily true it is not possible that *S* be false, and thus it is obviously not possible that simultaneously the premises be true and *S* false. (If it is not possible for the cow to jump over the moon, it is not possible for the cow to jump over the moon while the little dog laughs to see such sport.) Therefore, the argument is sound. On the other hand, if *S* is a consequence of *every* set of premises, it is a consequence of any set of premises consisting exclusively of necessary truths. Thus it will be true under all circumstances that make these sentences true, i.e., under all circumstances. Therefore, *S* is a necessary truth. So, *S* is necessarily true if and only if it is a consequence of every set of premises.

Example: the necessary conditional mentioned on page 7 is a consequence of

Grass is green,

for of course there are no conceivable circumstances of which this sentence would be true and the conditional false, since there are no circumstances of which the conditional would be false.

Since mathematical truths are considered to be the very paradigms of necessary truths, the considerations just raised lead us to realize that there is no such thing as an unsound argument having a mathematical truth as its conclusion. This shows that the concepts of 'sound argument' and 'proof' cannot very well be identified with one another; to find a proof for a mathematical sentence involves a great deal more than merely producing a sound argument that has only mathematical truths as premises and the given sentence as conclusion.

The above characterization of necessary truth is most often associated with the philosopher G. W. Leibniz (1646–1716), though it is by no means original with him. According to the doctrine of Leibniz, the actual world in which we find ourselves is only one of infinitely many possible worlds that could have existed. It is, indeed, the *best* of all possible worlds (for a reason not relevant here), and that is why God chose to bring it into existence. Now the distinction between the actual world and the various other possible worlds leads to a related distinction among sentences. To say that a sentence is true is to say that it is true of the actual world. Some true sentences, however, are true not only of the actual world but also of all other possible worlds as well; these are the necessary truths (sometimes called also 'truths of reason', or 'eternal truths'). Sentences that are true of the actual world but not of all possible worlds are called 'contingent truths' (or 'truths of fact'). Scientific laws are supposed to be of the latter kind; for, although they describe the actual world in considerable generality, there are numerous possible worlds of which they do not hold. The truths of mathematics and logic, on the other hand, are said to be true of all possible worlds, including of course the actual world. Thus, for example, there is no possible world of which the sentence 'Socrates died in 399 B.C. or Socrates did not die in 399 B.C.' would be false. (In considering whether this is so, it is well to remember that the question is: given the sense of this sentence, is there a possible world of which it would, in that sense, be false? The circumstance that it would be false if, e.g., the word 'or' meant what 'and' means, is irrelevant.)

Leibniz sometimes explains possibility and necessity in terms of 'conceivability': a sentence is a necessary truth if the opposite (of what it asserts) is inconceivable. But he wishes to use the latter word in what may be termed a 'non-psychological' sense; he is not interested in the sort of conceivability involved when we say that one person can conceive of what another finds inconceivable. So he explains conceivability as follows: a state of affairs is conceivable if no contradiction follows from the assumption that it exists. Of course we are here coming round full circle, from logical consequence to necessity to possibility to conceivability to logical consequence. In an informal explanation, however, such circularity is by no means disastrous, since a circular system of explanations may very possibly succeed in 'giving the idea' of what is to be explained, and in any case it can at least indicate certain interrelationships among the terms involved.

4. *Parenthetical remarks.* At this point it may be useful to insert a few comments about what we do mean and what we do not mean by some of the things we have said up to now. When we say that every declarative sentence is true or false we distinguish sharply between 'true' and 'known to be true'. If a sentence is known to be true, it is true; but many sentences are true although not known to be true. Thus from the fact that we do not know whether a given sentence is true or false it does not follow that the sentence *is* neither true nor false. In like manner we need to distinguish falsehood from oddity. Whether a sentence is true or false has no simple connection with whether in certain situations its utterance would be odd or misleading. For instance, if only one person has taken a certain examination and he has failed, it is probably misleading but clearly true to say 'all who have taken the examination have failed'. Nor is truth incompatible with what is sometimes called 'logical oddity'. It is supposed to be 'logically odd' for Smith to say:

He has gone out, but I don't believe it.

No one seems to deny, however, that someone else might truthfully say

He has gone out, but Smith doesn't believe it

or that Smith himself might later truthfully say (in reference to the very same situation)

He had gone out, but I didn't believe it.

This makes it evident that Smith might truthfully utter the first of these three sentences, too. Of course in most cases one either believes what one says or attempts to conceal the fact that one does not, and hence it is difficult (though not impossible) to think of circumstances in which a rational person would utter the 'logically odd' sentence in question. But that has very little to do with whether or not there are circumstances of which the given sentence would be *true*.

Concerning the term 'possible', it should be noted first of all that we are not using this term in the sense in which the impossible is only what American businessmen take a little longer to do. Instead we attempt (with perhaps questionable success) to distinguish technological, physical, and logical possibility. Something is technologically possible if, in the present state of technology, there is a way of doing it. It is physically possible if the hypothesis that it occurs is compatible with the (ideal) laws of nature. As technology advances, more and more things become technologically possible, but this advance always takes place within the domain of physical possibility. Many feats that are physically possible are now technologically impossible and perhaps always will be. Logical possibility is wider even than physical possibility. An event is logically possible if the hypothesis that it occurs is compatible with the laws of logic. If something is technologically impossible but physically possible, we may hope that eventually some clever man may find a way to get it done. If something is physically impossible, however, no businessman, engineer, or scientist, no matter how ingenious, will ever bring it off. If something is logically impossible, even the Deity joins this list. At least, that is how the story goes.

Another matter deserving explanation is our decision to take *sentences* as the objects with which logic deals. To some ears it sounds odd to say that sentences are true or false, and throughout the history of the subject there have been proposals to talk instead about statements, propositions, thoughts, or judgments. As described by their advocates, however, these latter items appear on sober consideration to share a rather serious drawback, which, to put it in the most severe manner, is this: they do not exist.

Even if they did, there are a number of considerations that would justify our operating with sentences anyway. A sentence, at least in its written form, is an object having a shape accessible to sensory perception, or, at worst, it is a set of such objects. Thus

It is raining

and

Es regnet,

though they may indeed be synonymous, are nonetheless a pair of easily distinguishable sentences. And in general we find that as long as we are dealing with sentences many of the properties in which the logician is interested are ascertainable by simple inspection. Only reasonably good eyesight, as contrasted with metaphysical acuity, is required to decide whether a sentence is simple or complex, affirmative or negative, or whether one sentence contains another as a part. Even so venerable and strange an issue as whether the conclusion of a sound argument is always somehow contained in the premises is not hard to settle if we take it as referring to sentences.

But matters are quite otherwise when we try to answer the same kinds of questions about propositions, statements, thoughts, and judgments. Propositions, we are told, are the senses or meanings of sentences. They are so-called abstract entities and, as such, are said to occupy no space, reflect no light, have no beginning or end, and so forth. At the same time, each proposition is regarded as having a *structure,* upon which its logical properties essentially depend. If we were to study logic from this point of view, therefore, it would be imperative to have a way of finding out in given cases what that structure is. Unfortunately no simple method is ever given. Since everyone agrees that sentences of wholly different shapes and structures may have the same meaning, the structures of propositions must not be confused with those of their corresponding sentences; yet a perusal of the literature leaves little doubt that just this sort of confusion frequently takes place. Sometimes, to be sure, one is advised not to rely upon sensory perception at all, but instead to look directly upon the proposition by means of the 'mind's eye'. Whoever admits that he is unable to do this leaves himself open to an obvious and uncomplimentary diagnosis traditionally employed in such cases. But, when all the rhetoric has been expended, we are left once again with the problem of how in practice to ascertain the structure of propositions expressed by given sentences.

Like difficulties are involved in talk about statements; these, though they purport to be different from propositions, clearly constitute another house of the same clan. Their partisans inform us that when we utter a sentence in certain favorable circumstances we succeed in 'making' a statement; the statement, not the sentence used, is 'properly' true or false. The same sentence, with the same meaning, may be used to make different statements. E.g., if I say

He won the election

while referring to Mr. Kennedy, I am making one statement; if I utter the same sentence while referring to Mr. Nixon, I am making another. Also, different sentences, with different meanings, may be used in making the same statement. For instance, if I say

Kennedy won the election,

I have made the same statement I made earlier when using 'he' to refer to Kennedy, but I have used a different sentence with a different meaning. Thus, as with propositions, the structure of a statement cannot be determined simply by looking at the sentence that is used in 'making' it. Yet the friends of statements do not hesitate to classify them as singular, general, or existential, as conjunctive, hypothetical, affirmative, negative, necessary, contingent, etc., or to say that they are of 'subject-predicate form', or that some of them 'have the same form' as others, or that they 'contain descriptions', and so on.

Similar remarks apply *mutatis mutandis* to thoughts and to judgments. Thoughts, in addition to sharing some of the ephemeral quality of propositions and statements, have another disadvantage as objects of logic. For, if logic is about thoughts, the laws of logic seem to become 'laws of thought', and the whole subject turns into a bit of out-dated psychology. Either the laws are presented as descriptions of how people do think (in which case they are false), or they are supposed to describe how people ought to think (in which case they are again false). Judgments, on the other hand, are perhaps the most obscure of all the candidates. According to one author (and it is hard to find two that agree) a judgment is 'an action of mind, and consists in comparing together two notions or ideas of objects derived from simple apprehension, so as to ascertain whether they agree or differ'. From this point of view logic is concerned with an 'act of mind called discourse or reasoning, by which from certain judgments we are enabled, without any new reference to the real objects, to form a new judgment.'* But surely the question whether, having performed a certain mental act, we are able to carry out another one, is (if it has any sense at all) a question of psychological fact; and thus once again all questions about the soundness or unsoundness of arguments would likewise become questions of fact, to be settled by studying the mental processes of human beings. Such a conception of logic is impossible to reconcile with the actual practice of logicians, including, of course, that of the judgment-logicians themselves.

In the present context the whole matter is not worth pursuing further, for in any case much of what we shall say about the logical properties of sentences can be carried over, in a fairly regular way, into talk about propositions, statements, thoughts, or judgments. In the main we shall

*Jevons, W. S., *Elementary Lessons in Logic,* London, 1909, pp. 12, 14.

avoid sentences like

It is raining

and

He is here,

which contain such 'egocentric' words as 'here', 'now', 'this', 'he', 'you', 'is' (in the sense of 'is now') and depend for their truth or falsity upon when, where, and by whom they are uttered. This sort of dependence can usually be obviated by using a sentence in which persons, times, and places are specified, e.g.,

It was raining in Berkeley, California, at 4 p.m. on December 28, 1961

or

President Kennedy was in Vienna on July 24, 1961.

By and large, if we confine ourselves to relatively unambiguous sentences not containing egocentric words, the following equivalences seem likely to be accepted: a sentence is true if and only if the proposition expressed by it is true; a sentence is true if and only if the statement that ordinarily would be made by using it is true. Thoughts and judgments may be treated analogously. If, therefore, the student has already acquired philosophic scruples against saying that sentences are true or false, he may utilize the equivalences given; otherwise, he may safely forget about the whole thing.

5. *Logical form.* One of the most obvious and yet most important facts about necessary truths was noticed by Aristotle, at the very beginning of the history of logic. It is this. In a large class of cases, when a sentence is necessarily true all sentences of what may be called 'the same logical form' are likewise necessarily true. The point will become more clear in examples. Consider the necessary truth

Socrates died in 399 B.C. or Socrates did not die in 399 B.C.

This sentence remains true and even necessary when its sentential component 'Socrates died in 399 B.C.' is replaced (at both of its occurrences) by any other sentence whatever. Thus the sentences

Aristotle died in 399 B.C. or Aristotle did not die in 399 B.C.,
Aristotle taught Alexander or Aristotle did not teach Alexander,
Plato taught Alexander or Plato did not teach Alexander,
Kennedy won the election or Kennedy did not win the election

are all necessary truths and so is every other sentence of the form

S or not S.

Again, the sentences

> If grass is green and snow is white, then snow is white,
> If roses are red and snow is white, then snow is white,
> If roses are red and grass is green, then grass is green,
> If Aristotle taught Alexander and Alexander was dull, then Alexander was dull

are necessary, as are all other sentences of the form

If *S* and *T*, then *T*.

Still other examples are afforded by the conditional corresponding to the first argument mentioned in section 1 above, and by all other conditionals of the form

If all *B* is *C* and all *A* is *B*, then all *A* is *C*.

When, as in each of these cases, a necessary truth has the property that every other sentence of the same form is also necessary, we shall say that it is *necessary by virtue of its logical form,* or *logically true.*

Lest anyone suppose that every necessary truth is necessary by virtue of its form, we hasten to add the timeworn example

No bachelor is married,

which is necessary although many other sentences of the same form, e.g.,

No Senator is married,

are not.

Even though not every necessary truth is necessary by virtue of its logical form, however, a plausible case has been made for the somewhat weaker claim that every necessary truth is either itself a sentence necessary by virtue of its form or else is obtainable from such a sentence by substituting synonyms for synonyms. The sentence

No bachelor is married,

for example, is obtainable from the sentence

No unmarried man is married,

i.e.,

No man who is not married, is married

by putting 'bachelor' for its synonym 'unmarried man'; and the last of these sentences is necessary by virtue of its form, since every sentence of the form

No *A* that is not *B*, is *B*

is a necessary truth.

We may note in passing that in view of the interdependence of soundness and necessity one can make a corresponding division of sound arguments into those that are sound by virtue of their logical form and those that are obtainable from such arguments by putting synonyms for synonyms. Thus, every argument of the form

All *B* is *C*;
All *A* is *B*;
Therefore, all *A* is *C*

is sound and will belong to the first type just distinguished. The argument

Smith is a bachelor;
Therefore, Smith is not married

is sound and belongs to the second type, for it is obtainable from an argument of the form

x is a *B* that is not *C*;
Therefore, *x* is not *C*

by putting 'bachelor' for 'man who is not married'.

Thus the concept of logical form is obviously of central importance for our subject; equally obviously it is a concept requiring a great deal of clarification. We need practical criteria for deciding what the logical form of a given sentence is, and which are the forms that have only necessary truths as instances. Unfortunately, the irregularity of natural languages makes the achievement of such criteria difficult if not altogether impossible; only in the case of artificial languages is there any real prospect of success.

Let us look for a moment at some of the expressions we have used in describing the forms of sentences:

S or not *S*
If *S* and *T*, then *T*
If all *B* is *C* and all *A* is *B*, then all *A* is *C*
No *A* that is not *B*, is *B*
x is a *B* that is not *C*

In the first two of these cases one obtains particular sentences by replacing the letters '*S*' and '*T*' by sentences. These letters, and all others to be used for the same purpose, may be called *sentential letters*. In the next two cases one must substitute general terms (e.g., 'men', 'mortals', 'professors', 'Swedes'), not sentences, in order to obtain particular sentences. The letters '*A*', '*B*', '*C*', and all others for which general terms are to be substituted, are *class letters*. In the last case we have used the letter '*x*', for which names of individuals, like 'Socrates', 'Aristotle', 'Alexander', or descriptions of individuals, like 'the teacher of Alexander', 'the student of Plato', are to be substituted. Letters for which expressions denoting individuals are to be substituted are called *individual letters*. And, while we are at it, let us introduce the term *matrix* to refer to the formal expressions themselves: a matrix is an expression built up out of so-called *logical words* (like 'and', 'or', 'if . . . then', 'not', 'all', 'is', etc.), together with sentential, class, or individual letters, and such that the result of replacing the letters by the appropriate kinds of expressions is a sentence.

Thus we may say that a sentence is a truth of logic if and only if it is a substitution-instance of a matrix all instances of which are necessary truths. Observe that the notion of matrix, which here plays a crucial role, will vary with the choice of a list of logical words; unfortunately the question as to which words should be considered logical and which not, involves a certain amount of arbitrariness.

Sometimes one defines the truths of logic as substitution-instances of matrices all instances of which are *true*. The advantage of this approach is that it explains logical truth in terms of truth, which is a relatively clear notion. Its principal disadvantage is its presupposition that the language contains a name or description for every object and a corresponding predicate for every property. (Otherwise the nameless objects or properties might be just the objects or properties for which a given matrix would be false). But for various reasons this supposition is not plausible.

The above idea has an important variant, however, that escapes the difficulty mentioned. We may define a sentence as a truth of logic if it is a substitution-instance of a matrix that 'comes out true' no matter how one interprets the various individual, class, and sentential letters occurring therein; where interpreting a letter consists in assigning an appropriate object (individual, class, or truth-value) to it as its denotation. Thus, the matrix

$$x \text{ is an } A \text{ or } x \text{ is not an } A$$

comes out true no matter what individual is assigned to 'x' and what class of individuals is assigned to 'A', and hence every instance of it is a truth of logic. This approach, which seeks to explicate logical truth in terms of a notion of 'coming out true with respect to an interpretation', is in effect the one to be used in our later developments.

6. *Artificial languages*. In recent years nearly all serious study of logic has employed the device of an 'artificial' or 'formalized' language. The idea of constructing such languages goes back at least as far as Leibniz, although the first really successful attempt was made by the German logician Gottlob Frege, in his *Begriffsschrift* (1879).

It has always been clear that *some* degree of formalization is inevitable if the subject is to be pursued at all. Even Aristotle had to introduce a certain stylization of language in order to carry out his investigations, for the very simplest of what have been called 'laws of formal logic' always require some adjustments in their application to a natural (written) language. For instance, although the sentence

All men are mortal or not all men are mortal

is an exact substitution-instance of the matrix

S or not *S*,

i.e., can be obtained from it by putting 'all men are mortal' for the letter '*S*' (though even here we must make a slight adjustment by capitalizing the initial 'a' in one of its occurrences and not in the other), the sentence

Socrates is mortal or Socrates is not mortal

is not strictly such an instance. In connection with this particular matrix the difficulty stems from the fact that, although the denial of an English sentence is usually formed by adding the word 'not', there is no simple rule for determining *where* the addition is to be made. We form the denial of 'Socrates is mortal' by inserting the word 'not' right after the verb, but if we try to deny the sentence 'all men are mortal' in the same way, the result is

All men are not mortal,

whereas what is wanted is the much weaker sentence

Not all men are mortal.

To deny the sentence

Smith cracked the safe and Jones drove the getaway car

one is forced to prefix some such phrase as 'it is not the case that . . .', or possibly even to substitute

Either Smith did not crack the safe or Jones did not drive
the getaway car.

(This example also shows that even so simple a sentence as 'Smith cracked the safe' cannot be denied merely by adding 'not'; two other changes must also be made).

Thus, although when we say that all sentences of the form

S or not *S*

are logically true we mean to include such cases as

All men are mortal or not all men are mortal,
Socrates is mortal or Socrates is not mortal,
Smith cracked the safe or Smith did not crack the safe,

> Either Smith cracked the safe and Jones drove the getaway car, or it is not the case that Smith cracked the safe and Jones drove the getaway car,

we see that, strictly speaking, relatively few of these are of the form mentioned. Similarly, the sentence

> If all men are mortal and all Greeks are men, then all Greeks are mortal

is not strictly of the form

If all B is C and all A is B, then all A is C,

since 'is' has to be adjusted everywhere to 'are'. Adjustments of this sort are the rule rather than the exception. In general, it is clear that for the natural language there are few, if any, matrices that *literally* have only necessary truths as substitution-instances.

If, therefore, we try to develop a formal logic of the natural language, we find ourselves unable to proceed in an exact way, even in the very simplest cases. An obvious remedy for some of these difficulties, and a remedy utilized by nearly all modern logicians, is to set up an artificial language that is grammatically simpler and more regular than the natural language. For such a language one can hope to characterize the notion of logical truth in a satisfactorily precise way. After another chapter of preliminaries we shall follow this approach.

EXERCISES

1. Try to classify the following sentences as necessarily true, contingently true, or false. Of the necessary truths indicate which are logically true and which are not. Discuss any cases in which ambiguity or vagueness seems crucial.
   (a) The population of Berkeley is greater than 100,000.
   (b) Some concert pianists are Frenchmen.
   (c) Every black dog is a dog.
   (d) Every suspected criminal is a criminal.
   (e) Every suspected criminal is suspected.
   (f) All ravens are black.
   (g) If two people are brothers they are siblings.
   (h) If one man is taller than a second, then the second is shorter than the first.
   (i) Whatever goes up must come down.
   (j) Two bodies cannot be in the same place at the same time.
   (k) In English, 'England' denotes England.
   (l) For all positive integers x, y, z: if $x < y$, then $x + z < y + z$.

2. The following argument was offered in the Middle Ages as an example to show that *every* sentence follows from a contradiction, and indeed follows *formally* from it.

(1) Socrates exists and Socrates does not exist.	Premise
(2) Socrates exists.	From (1)
(3) Socrates exists or the stick is standing in the corner.	From (2)
(4) Socrates does not exist.	From (1)
(5) The stick is standing in the corner.	From (3) and (4).

Does the argument seem intuitively sound? If not, at what point does it appear to break down? Can you construct an analogous argument deriving 'Either Socrates exists or Socrates does not exist' from 'the stick is not standing in the corner'?

3. Give an example (if possible) of:
   (a) a matrix of which all substitution-instances are necessary truths;
   (b) a matrix of which some substitution-instances are necessary truths and some are not;
   (c) a matrix of which no substitution-instance is a necessary truth;
   (d) two matrices such that all substitution-instances of the first are substitution-instances of the second, but not vice versa;
   (e) a sound argument with true premises and false conclusion;
   (f) an unsound argument with '$2 + 2 = 4$' as conclusion;
   (g) an argument that has '$2 + 2 = 4$' as conclusion and is not formally sound;
   (h) a sound one-premised argument that remains sound when premise and conclusion are replaced by their respective denials;
   (i) two arguments, one unsound and one sound, and each of which has one false premise, one true premise, and a true conclusion.
4. Which of the following generalizations are correct?
   (a) For any three sentences *R*, *S*, *T*: if *R* follows from *S* and *S* follows from *T*, then *R* follows from *T*.
   (b) For any two sentences *R*, *S*: if *R* follows from *S*, then *S* follows from *R*.
   (c) For any two sentences *R*, *S*: if *R* follows from *S*, then *not S* follows from *not R*.
   (d) For any three sentences *R*, *S*, *T*: if *R* follows from *S* and *T*, then *not T* follows from *S* and *not R*.
   (e) For any three sentences *R*, *S*, *T*: if *R* follows from *S* and *T*, then *if S then R* follows from *T* alone.
   (f) For any three sentences *R*, *S*, *T*: if *R* follows from *S* and also from *not S*, then *R* follows from *T*.

2

FURTHER PRELIMINARIES

This chapter is devoted to a miscellany of topics that require attention before we embark upon our study of formalized languages. The first of these topics is the so-called *use-mention* distinction—the distinction between *using* a linguistic expression to refer to something else and *mentioning* the expression (i.e., talking about it). Secondly, we set forth Frege's distinction between the sense and denotation of linguistic expressions, noting in particular his view that in certain contexts an expression denotes what is ordinarily its sense. The third topic is that of *variables;* since these play a prominent role in our subsequent discussion, it is especially important that our use of them be thoroughly understood. Variables are simply letters of the alphabet; their *substituends* are the names or other expressions that may meaningfully be substituted for them, and their *values* include all objects named by substituends. In the fourth section we explain the notion of *sentence-form.* Roughly, a sentence-form is an expression that is either a sentence or may be obtained from a sentence by replacing some or all occurrences of names by variables. The notion of sentence-form is helpful in connection with our fifth topic, *sets.* Since the predicates of the artificial language will be interpreted as referring to sets and relations, it is necessary at this point to introduce a brief, intuitive characterization of these entities. Lastly we draw attention to the distinction between *object-language* and *metalanguage,* i.e., between the language *about* which we talk (in a given discussion) and the language *in* which we talk.

1. *Use and mention.* The use-mention distinction is absolutely essential for understanding modern treatments of logic. One must be able to distinguish sharply between cases in which a given linguistic expression is *mentioned,* i.e., spoken about, and those in which it is *used* (perhaps to mention something else). This involves making an equally sharp distinction between names and what they name.

To speak about or mention an object we ordinarily use a linguistic expression that names or in some other way refers to the object. As a rule the name or other expression used is not the same as the thing referred to, which is mentioned by using the name. When speaking about the Campanile, for example, we of course make use of the *word* 'Campanile' and not the Campanile itself.

In most situations there is no danger at all of confusing the name with the thing named, although trouble does often occur when the thing named is an abstract entity, not perceivable by the senses. Thus, while no one has difficulty in distinguishing the Campanile from the word 'Campanile', there has been some tendency to confuse numbers, which (according to received doctrine) are not to be found in sensory experience, with the numerals and other numerical expressions that purport to denote them. It is tempting to believe, for instance, that 12 and $7 + 5$ are two closely related (equal) but nonetheless different (i.e., not identical) numbers.

Care is required, too, when the things to be mentioned are themselves linguistic expressions. In speaking of a word or other linguistic expression we shall ordinarily follow the same rule that applies when the object to be spoken about is something extralinguistic: use a name or description of the object and not just the object itself. One of the most convenient ways of forming a name of a given linguistic expression is that of placing the expression within quotation marks. For example,

'The Iliad'

denotes the title

The Iliad,

which in turn denotes the great epic poem. If we wish to speak of the title, we use a name of the title and not just the title itself; thus we may say

'The Iliad' is a short title.

According to the rule stated above, omission of the quotation marks here would result in our talking about the Iliad rather than about its name.

Another useful device for naming linguistic expressions is that of the displayed line, three instances of which may be seen in the preceding paragraph. As will already have become obvious, this device is frequently employed in the present book.

In ordinary discourse it is natural to rely upon context and good sense to tell us whether an expression is being used or mentioned. If we read

Shakespeare's real name was Bacon,

we take this in the sense of

Shakespeare's real name was 'Bacon',

since the other interpretation is absurd. In the study of logic, however, it is advantageous to be rather strict in following the convention that to mention an object one shall use a name or description of that object, and not just the object itself. The word 'just' is inserted here in order to take account of those few cases in which a name or description of the object contains the object as a part, as for example in the expression 'Bertrand Russell's name'.

2. *Sense and denotation.* We have been drawing attention to the distinction between a linguistic expression and the object, if any, that it denotes. A further distinction, philosophically more questionable but certainly very useful, is that drawn by Gottlob Frege between the sense and denotation (*Sinn* and *Bedeutung*) of linguistic expressions. Under the heading of 'linguistic expressions' in this connection, he includes names, descriptions, predicate expressions, and full sentences. According to Frege, the sense of such an expression is its meaning; one must 'grasp' the sense if one is to understand the expression when it is used. (In a similar vein the ancient Stoics described the sense of Greek discourse as 'what the Greeks grasp, but the barbarians do not, when Greek is spoken'.) The denotation, on the other hand, is the object or objects, if any, to which the expression refers. Thus 'the Morning Star' and 'the Evening Star' have the same denotation, namely the planet Venus, but they have different senses. The former expression, says Frege, has approximately the sense of 'the brightly shining object that appears in the morning in the eastern sky', while the latter does not. Given the sense of an expression, the denotation is uniquely determined, but, as in the above example, different senses may correspond to, or determine, the same denotation.

In Frege's view, sense and denotation must satisfy two further conditions:

a) The denotation of a complex expression is a function of the denotations of its parts. In other words, if a part of a complex expression is replaced by another expression having the same denotation as that part, the denotation of the whole is not altered thereby.

b) Similarly, the sense of a complex expression is a function of the senses of its parts.

These two principles, together with some rather obvious facts about the meanings of particular expressions, serve to establish the point that sense, or meaning, cannot plausibly be identified with denotation. For the two sentences

(1) The Morning Star is the same as the Evening Star,
(2) The Morning Star is the same as the Morning Star

obviously do not have the same sense, since the second expresses a triviality while the first does not. Yet, as noted above, the expressions 'the Morning Star' and 'the Evening Star' do have the same denotation; if sense were the same as denotation, they would have the same sense. Thus, by principle *b*), they would be interchangeable in (1) without altering its sense, so that (1) would have the same sense as (2). Or, on the same basis, we can apply principle *a*) to obtain the result that (1) and (2) have the same denotation, so that again denotation cannot be identified with sense.

This last consideration involves the idea of the denotation of a sentence. According to Frege, when a sentence stands alone or when it stands as a co-ordinate clause in a larger sentence, it denotes its truth-value and has as its sense a so-called proposition or thought. Thus (1) and (2) have the same denotation, although they differ in sense.

(In view of the doubts earlier expressed about the existence of such things as propositions—and the same doubts apply almost as well to truth-values—the reader may wonder how we dare to invoke these entities in the present context. The answer is that the essentials of Frege's view can be stated without any metaphysical assumptions about the existence of such things as propositions. For instance, it is possible to rephrase principle *b*) as follows: if expressions *S* and *T* are synonymous and expressions *U* and *V* are alike except that *U* contains an occurrence of *S* where *V* contains an occurrence of *T*, then *U* is synonymous with *V*. Since the nonexistence of propositions and the various other nonentities in no way requires us to deny that there are indeed pairs of synonymous expressions, we can accept most of Frege's semantics without agreeing to his metaphysics. The same charitable reading may of course be extended to the use of 'proposition' by other authors, insofar as such use is eliminable in like manner.)

It might be thought that Frege's theory could be confronted with a counterexample in the following way. Consider the sentences:

(3) Somebody wonders whether the Morning Star is the same as the Evening Star,
(4) Somebody wonders whether the Morning Star is the same as the Morning Star.

It seems reasonable to suppose that (3) is true and (4) false, i.e., that (3) and (4) have different denotations, even though principle *a*) seems to imply the contrary. Frege handles this sort of difficulty by denying that a given expression always has the same sense in whatever context it may occur. In fact, he says, in contexts governed by such phrases as 'believes that', 'knows that', 'wonders whether', etc., a linguistic expression denotes what is ordinarily its sense, and has as sense something else. Thus when sentence (1) stands alone it denotes its truth-value and expresses (has as sense) a certain proposition, but when it occurs in (3) it denotes that proposition. Correspondingly, the expression 'Morning Star', when it occurs in (1) standing alone, denotes the planet Venus; but at its occurrence in (3) it denotes the sense it has in (1). Note also that since its denotation in (3) is different from its denotation in (1) and sense uniquely determines denotation, its sense in (3) is different from its sense in (1).

Thus Frege distinguishes the *ordinary* or *direct* sense and denotation of an expression from its *oblique* or *indirect* sense and denotation, specifying that the oblique denotation is the same as the ordinary sense. For future reference we define an occurrence of a name or description in a given expression as a *direct occurrence* if in that context the name or description has its ordinary sense (and denotation, if any); otherwise, the occurrence is *indirect*. We shall return to this topic in section 4.

3. *Variables.* The use of variables goes back to the time of Aristotle. Today they are found in all types of technical and non-technical discourse, from mathematics to Gilbert and Sullivan, inclusive.

> See how the Fates their gifts allot,
> For *A* is happy—*B* is not.
> Yet *B* is worthy, I dare say,
> Of more prosperity than *A*.
> (*The Mikado*)

Variables are best regarded simply as letters of the alphabet and not as things that vary (or as names of things that vary). They are used for a number of purposes, one of the most important of which is to facilitate the expression of generalizations. Thus, if a jurist wishes to discuss all cases of a certain type, he may employ letters instead of names of particular persons; he may say, for instance:

> *A* signs a contract with *B*, who later dies, leaving *C* as his only heir. After waiting one year for *A* to carry out his obligation under the contract, *C* decides to sue. Meanwhile *A* has petitioned for bankruptcy, and . . .

This does not mean, of course, that there are three unknown persons, *A*, *B*, *C*, who are here under discussion and whose identity might later be discovered. Nor does it mean that there are three *variable* persons, A, B, C, who are involved in complicated dealings. It means only that if *any* three persons, known or unknown, variable or constant, are related as described, then they will stand in some other relation that the statement goes on to set forth.

Again, suppose that a mathematician wishes to state a general principle of which

$$6(4 + 3) = 6 \cdot 4 + 6 \cdot 3$$

and

$$8(1 + 9) = 8 \cdot 1 + 8 \cdot 9$$

are particular instances. He may write

$$x(y + z) = x \cdot y + x \cdot z$$

If we try to express the same law without the help of variables or some similar device, the result is cumbersome and unclear:

> Given any three numbers, not necessarily distinct, the first times the sum of the second and the third is equal to the sum of the first times the second and the first times the third.

As indicated earlier, variables have other uses besides the expression of generality, but for present purposes these do not need to be discussed. There are two questions, however, that one should be ready to answer whenever a variable is to be employed, namely, what are its *substituends* and what are its *values* (where the substituends are all those expressions that may be meaningfully substituted for the variable, and the values include all objects named by substituends). In

> If x is elected President, then x will serve for four years

the substituends for the variable 'x' would most naturally include names of people (e.g., 'Goldwater', 'Stevenson', 'Kennedy', 'Rockefeller') and would not usually be thought to include names of numbers (e.g., '6', '3 + 4'). The corresponding values would be people and not numbers. On the other hand, in

$$x + y = y + x$$

the substituends for 'x' and 'y' would normally be numerals and other numerical expressions, and the values of 'x' and 'y' would be numbers. These are matters of convention, of course, and the conventions vary from author to author and from context to context.

In our forthcoming descriptions of artificial languages we shall need to refer to certain symbols, as well as to formulas and other expressions composed of such symbols. When reference is to be made to a particular symbol or expression, its quotation-name will usually serve the purpose well enough; e.g., we can say 'v' is a logical constant.

But sometimes there is need to speak generally about all expressions of a certain form, and in such cases it is convenient to use variables. For instance, suppose that we wish to express compactly the following fact:

> For any two formulas, not necessarily distinct, the result of writing a left parenthesis, followed by the first formula, followed by the symbol 'v', followed by the second formula, followed by a right parenthesis, is again a formula.

With variables we can state this as:

(1) If ϕ and ψ are formulas, then $(\phi \vee \psi)$ is a formula.

(Here the substituends for 'ϕ' and 'ψ' are names of expressions of the artificial language under consideration, and the values are accordingly the expressions themselves.)

Thus, as a particular instance of (1) we should like to have

(2) If 'F^1a' and 'G^1b' are formulas, then '$(F^1a \vee G^1b)$' is a formula.

It will be noted, however, that this particular instance cannot strictly be obtained from (1) by substituting ''F^1a'' for 'ϕ' and ''G^1b'' for 'ψ' throughout. Such substitution would only yield

(3) If 'F^1a' and 'G^1b' are formulas, then ('F^1a' v 'G^1b') is a formula,

which is nonsense. To obtain the desired result we shall adopt (for later reference) two additional conventions: (i) when logical constants of our artificial language (e.g., symbols like 'v' or parentheses) appear in combination with names or variables referring to expressions of that language, these constants shall be considered names of themselves; (ii) to refer to a compound expression we juxtapose names or variables referring to its parts. On this basis, (1) is short for

(1′) If ϕ and ψ are formulas, then the result of writing '(', followed by ϕ, followed by 'v', followed by ψ, followed by ')' is again a formula.

Straightforward substitution of ''F^1a'' for 'ϕ' and ''G^1b'' for 'ψ' in (1′) gives

(2′) If 'F^1a' and 'G^1b' are formulas, then the result of writing '(', followed by 'F^1a', followed by 'v', followed by 'G^1b', followed by ')' is again a formula.

This may be rephrased as (2).

The student will not grasp the full import of these conventions until later, when he has become acquainted with our artificial languages and with the ways in which we use variables to describe them; at that time he may wish to reread the foregoing section.

4. *Sentence-forms.* Returning now to less technical matters, we attend once again to the distinction between direct and indirect occurrences of names or descriptions in sentences or other linguistic expressions. Consider the following argument:

> The man who committed the crime was carrying a gun.
> Smith is the man who committed the crime.
> Therefore, Smith was carrying a gun.

This argument is sound, for if the second premise is true, then the name 'Smith' and the description 'the man who committed the crime' denote the same person, and thus the conclusion and the first premise say exactly the same thing about the same person. But now consider the argument:

> Jones knows that the man who committed the crime was carrying a gun.
> Smith is the man who committed the crime.
> Therefore, Jones knows that Smith was carrying a gun.

It is easy to see that this argument, unlike the other one, is not sound. For even if the premises are true, Jones may not *know* that Smith is the man who committed the crime, and so there is little reason to conclude that Jones knows that Smith was carrying a gun.

The difference between these two cases is symptomatic of the fact that words following such expressions as 'knows that', 'believes that', 'wonders whether', 'necessarily', and the like, do not in general obey the same logical principles as they do when they stand outside the scope of such expressions.

When a name or description occurs directly in a sentence, its replacement by any other name or description having the same ordinary denotation will not alter the truth-value of the sentence. This is because the denotation of a sentence is its truth-value, and thus the Fregean principle that the denotation of a complex expression is a function of the denotations of its parts ensures that the truth-value will not be changed by replacements of the kind mentioned. When, on the other hand, the name or description occurs indirectly, its replacement by another expression having the same ordinary denotation *may* alter the truth-value, for in an indirect occurrence the name or description denotes its *oblique* denotation, i.e., its ordinary sense. Hence in such cases the truth-value of the sentence depends upon the ordinary sense, and not merely upon the ordinary denotation, of the expression being replaced. In the sentence

> The man who committed the crime was carrying a gun

the occurrence of the description 'the man who committed the crime' is direct, and replacement of this description by any other name or description of the same man—e.g., by 'Smith'—will not change the truth-value. But the occurrence of the same description in

Jones knows that the man who committed the crime was
carrying a gun

is not direct, and hence we cannot depend upon the truth-value of the sentence being maintained when another name or description of the same man is inserted.

Note that sentences like

'Mark Twain' is a pseudonym

present a somewhat different problem. We shall say not merely that the words 'Mark Twain' do not occur *directly* here, but that they do not occur at all. In general, an expression consisting of another expression surrounded by quotation marks is to be thought of as a unit, so that

Mark Twain

does not occur in

'Mark Twain'

any more than 'Art' occurs in 'General MacArthur' or 'men' occurs in 'Samuel Clemens'. On the other hand, the quotation name

'Mark Twain'

does of course occur directly in the sentence mentioned above, for as long as we replace it by any other name or description of the same name, the truth-value cannot be altered. Thus, if we replace it by the description 'The name under which Samuel Clemens published his books' we get

The name under which Samuel Clemens published his books
is a pseudonym,

which has the same truth-value as the sentence in which the replacement was made.

The distinction between direct and indirect occurrences of names is needed in order to define the notion of a sentence-form, a notion utilized in our subsequent discussion of sets. A *sentence-form* is an expression that is a sentence or is obtainable from a sentence by replacing some or all direct occurrences of names by variables. Thus

x was carrying a gun,
Smith was carrying a gun,
x is y,
Smith is y,
x knows that the man who committed
the crime was carrying a gun

are all sentence-forms, while

x knows that y was carrying a gun,
x believes that y is z,
Jones knows that Smith is z,
It is impossible that x is y.

are not.

Consider now such a sentence-form as

$$x + 6 < 8.$$

There are basically two ways in which a sentence (of 'logicians' English') may be obtained from this expression. We may replace the occurrence of 'x' by a name or description of a number, thus producing such sentences as

$$1 + 6 < 8, \qquad (4 + 10) + 6 < 8,$$
$$2 + 6 < 8, \qquad 2^{10} + 6 < 8.$$
$$3 + 6 < 8,$$

Or, we may prefix a so-called *quantifier*. Quantifiers are phrases like

For every x, . . .

and

There is an x such that . . .

The former is called a *universal quantifier,* and the latter an *existential quantifier.* Using this method we obtain the false sentence

For every x, $x + 6 < 8$

and the true sentence

There is an x such that $x + 6 < 8$.

We shall extend the use of quantifiers, which are common in mathematical discourse, to discourse of all types. For instance, we shall understand

There is an x such that x is now President of the U. S. A.

to be a true sentence asserting that at least one person is now President of the U. S. A., while the sentence

For every x, x is now President of the U. S. A.

will be false, stating as it does that everybody and everything is President. Note that in the definition of 'sentence-form' given on the preceding page, the term 'sentence' is to be understood to cover sentences which, like those just mentioned, contain quantifiers as well as variables to which those quantifiers apply.

Sometimes combinations of quantifiers are used, as in

For every x there is a y such that $x \neq y$,

which is true, and

There is a y such that for every x, $x \neq y$,

which is false. The former of these says truly that whatever object is chosen, there exists something different from it; the latter says that some object is different from everything (including itself).

Strictly speaking, a sentence-form that is not a sentence has no truth-value. Nevertheless, such expressions are often asserted. In many cases it is evident that the given expression is to be taken as implicitly preceded by universal quantifiers binding the relevant variables, so that

$$x \cdot (y + z) = x \cdot y + x \cdot z$$

represents

For all numbers x, y, z: $x \cdot (y + z) = x \cdot y + x \cdot z$.

This is not always so, however. For example, the sentence-form

$$x^2 - 4 = 0$$

is not usually employed as a short way of writing the false sentence

For every number x, $x^2 - 4 = 0$.

But a full explication of this point would take us too far afield.

In the sentence-form

$$x < 6$$

it is necessary either to substitute something for 'x' or else to prefix an operator like 'for every x' or 'there is an x such that' in order to obtain a sentence. In the sentence-form

There is an x such that $x < y$,

it is similarly necessary to substitute for or quantify upon 'y' if a sentence is to be obtained, but note that the two occurrences of 'x' may be left unchanged. Thus

There is an x such that $x < 6$

is a sentence, and so is

For every y there is an x such that $x < y$.

From these examples we see that the occurrences of variables in a sentence-form may be divided into two kinds: (1) those requiring replacement or quantification in order to obtain a sentence, and (2) the others. The former are called *free* occurrences, the latter *bound.*

For example, in the sentence-form

There is an x such that $x < y$,

the two occurrences of 'x' are bound, and the single occurrence of 'y' is free. In

For every x and for every y, $x + y = y + x$,

which of course is a sentence as it stands, all occurrences of variables are bound. In

$x < 6$ and, for every x, $x < y$

the first occurrence of 'x' and the only occurrence of 'y' are free; the remaining two occurrences of 'x' are bound. For by putting numerals in place of the free occurrences we obtain, e.g., the sentence

$5 < 6$ and, for every x, $x < 1$.

Note that according to our use of 'occur' it turns out that 'x' does not occur at all in the sentence-form ''x' is a letter of the alphabet' and hence occurs neither free nor bound.

Now suppose that only one variable, e.g., 'x', occurs free in a given sentence-form. Then we shall say that a given object *satisfies* that sentence-form if the result of interpreting the free occurrences of the variable as referring to that object is true. Thus the sentence-form

$4 < x$ and $x < 9$

is satisfied by the numbers 5, 6, 7, and 8, but not by the numbers 3 or 10. Similarly, Senators Morse and Douglas satisfy the form

x was a member of the U. S. Senate in 1962,

while Messrs. Stevenson and Nixon do not. In the case of a form in which exactly two variables occur free (though of course each may occur several times), we shall say that ordered *pairs* of objects satisfy or fail to satisfy the form. In this connection the variables are taken in alphabetical order. For instance, the pair consisting of the integers 3 and 6, in that order, satisfies the form

$x < y$,

while the opposite pair, consisting of 6 and 3, does not. The latter pair, but not the former, satisfies the form

$y < x$.

When three variables occur free we must consider ordered *triples* of objects. The triple consisting of 3, 5, 7 satisfies the form

$x < y$ and $y < z$,

since

$3 < 5$ and $5 < 7$

is true; but the triple consisting of 3, 5, and 3 does not, for

$$3 < 5 \text{ and } 5 < 3$$

is false.

Another useful notion is that of a *description-form.* Description-forms are to descriptions as sentence-forms are to sentences. Thus

the teacher of x

becomes the description

the teacher of Alexander the Great

when the occurrence of 'x' is replaced by the name 'Alexander the Great'. As in the case of sentences, so with descriptions the distinction between direct and indirect occurrences of component names or descriptions is important. For only when the component occurs directly can we be sure that its replacement by another expression with the same ordinary denotation will not alter the denotation of the description in which the replacement is made. Thus, the description just quoted continues to denote Aristotle if we replace its component 'Alexander the Great' by any other name or description of the same person, e.g., by 'the son and successor of Philip', thus obtaining the description

the teacher of the son and successor of Philip.

On the other hand, the description

the British monarch who wished to know whether Scott was the author of *Waverley,*

which denotes George IV, turns into

the British monarch who wished to know whether Scott was Scott,

which does not denote George IV, when the occurrence of 'the author of *Waverley*' is replaced by another expression, viz., 'Scott', which refers to the same person. Therefore, just as we noted in the case of sentences, words following the phrases 'knows that', 'believes that', 'wonders whether', and many others, behave quite differently from those occurring directly. With this in mind we define a *description-form* as an expression that is either a description or obtainable from a description by replacing some or all direct occurrences of names by variables.

Examples of description-forms:

the son of x
the queen of x
$x + y$
x^2
the only child of x and y

Examples of expressions which, appearances to the contrary, are not description-forms:

> the person who is believed to have seen x
> the witness who testified that x was present
> the known accomplice of x
> the important theorem that x is believed to have proved

5. *Sets.* The concept of *set,* or *class,* is so basic that when asked to define it we can hardly do better than give synonyms and examples. Among the synonyms are the terms 'collection', 'aggregate', 'totality' (French 'ensemble', German 'Menge'). Thus we can say that any collection of objects constitutes a set, but perhaps this is no more informative than if we said that any set of objects constitutes a collection. None of the synonyms is quite right. 'Collection', for instance, connotes 'to collect' ('to bring together'), but it is in no way essential that the members of a set should be located in the vicinity of one another. Perhaps the only satisfactory method of characterizing sets is by means of axioms, but for the present a rough, intuitive account will have to suffice.

The objects that constitute a set are called its *elements* or *members.* Each set is uniquely determined by its members; in other words, sets having the same members are identical. It is customary to use the symbol '$\in$' for the phrase 'is a member (or element) of', so that if, for example, the letter 'P' denotes the set of prime numbers, the sentence

$$7 \in P$$

means that seven is a member of the set of prime numbers—in other words, that seven is a prime number.

Sets are closely related to sentence-forms of one free variable; with certain qualifications we may say that for each sentence-form of one free variable there is a set whose members are just those objects that satisfy the form. Thus, corresponding to the sentence-form

$$x \text{ is a man}$$

there is the set of men; corresponding to the form

$$x \text{ is an even integer and } x \text{ is divisible by } 3$$

there is the set of all even integers that are divisible by three. Not every set, however, is associated with a sentence-form in this way, since it can be shown that there are more sets than sentence-forms.

Let us consider some further examples of sets. There is a set having as members all objects that satisfy the sentence-form

$$x \text{ is different from } x.$$

Obviously this set has no members; it is called the *empty set* and will be denoted by 'Λ'. Since sets are uniquely determined by their membership, there is only one such set and we are justified in speaking of *the* empty set. Correspondingly the *universal set,* symbolized by 'V', is the set of all objects satisfying the sentence-form

$$x \text{ is identical with } x.$$

Further, given a name N of an object, we can construct a corresponding sentence-form by writing N in the blank of the expression

$$x \text{ is identical with } \_\_ ;$$

the set of all objects satisfying this form will have exactly one member, viz., the object denoted by N. Thus, for the sentence-form

$$x \text{ is identical with Socrates}$$

there is the set whose only member is Socrates; this set is denoted by

$$\{\text{Socrates}\}.$$

Similarly, if N and N' are names of objects, the set of all objects satisfying the sentence-form that results from putting N and N', respectively, in the first and second blanks of the expression

$$x \text{ is identical with } \_\_ \text{ or } x \text{ is identical with } \_\_$$

is a set having at most two objects as elements and is called a *pair* or *couple*. If the given names are 'Socrates' and 'Plato', the couple will be denoted by

$$\{\text{Socrates, Plato}\}$$

or equally well by

$$\{\text{Plato, Socrates}\}.$$

In general, the notation

$$\{x\}$$

will stand for the description-form 'the set whose only member is x'; similarly, the notation

$$\{x, y\}$$

will stand for the description-form 'the set whose only members are x and y', and so on, for any finite number of occurrences of variables, names, or descriptions. Thus

$$\{6, 8\}$$

denotes the set whose only members are the numbers six and eight, and

$$\{6, 8, 2, 6\}$$

denotes the set whose only elements are the numbers six, eight, two, and six—which of course is the same as the set whose only elements are the numbers six, eight, and two. Thus we have

$$\{6, 8, 2, 6\} = \{6, 8, 2\}.$$

Next, we consider some relationships among sets. If every element of a set A is also an element of a set B, i.e., no element of A fails to be an element of B, the set A is said to be *included in,* or to be a *subset of,* the set B; in symbols*

$$A \subset B.$$

From the meaning of 'included in' it follows that, if A, B, and C are any sets whatever, we have:

(1) $\Lambda \subset A$
(2) $A \subset A$
(3) $A \subset \mathrm{V}$
(4) If $A \subset B$ and $B \subset A$, then $A = B$
(5) If $A \subset B$ and $B \subset C$, then $A \subset C$

Sets may be combined in certain ways to form new sets. If A and B are sets, then the *union* of A and B, in symbols

$$A \cup B,$$

is that set which has as its members all objects belonging to A or to B or to both. Similarly, the *intersection* of A and B, in symbols

$$A \cap B,$$

is that set which has as its members all objects belonging both to A and to B. And the *complement* of A, in symbols

$$A',$$

is the set of all objects not belonging to A. In terms of these operations we may formulate some further laws about sets. Thus, if again A, B, C are any sets whatever, we have:

(6) $A \cup B = B \cup A$
(7) $A \cap B = B \cap A$
(8) $A \subset A \cup B$
(9) $A \cap B \subset A$
(10) $\mathrm{V}' = \Lambda$
(11) $\Lambda' = \mathrm{V}$
(12) $A \cup \Lambda = A$
(13) $A \cup \mathrm{V} = \mathrm{V}$
(14) $A \cap \Lambda = \Lambda$
(15) $A \cap \mathrm{V} = A$
(16) $A \cup A = A$
(17) $A \cap A = A$
(18) $A \cup A' = \mathrm{V}$
(19) $A \cap A' = \Lambda$
(20) $A'' = A$
(21) $A \cup (B \cup C) = (A \cup B) \cup C$

* Note that the symbol here introduced is frequently used in the sense of 'is included in but not identical with'.

(22) $A \cap (B \cap C) = (A \cap B) \cap C$
(23) $A \cup (B \cap C) = (A \cup B) \cap (A \cup C)$
(24) $A \cap (B \cup C) = (A \cap B) \cup (A \cap C)$
(25) $A \cup (A \cap B) = A$
(26) $A \cap (A \cup B) = A$
(27) $(A \cup B)' = A' \cap B'$
(28) $(A \cap B)' = A' \cup B'$
(29) $A \subset B$ if and only if $A \cup B = B$
(30) If $A = B'$, then $B = A'$
(31) If $A \subset C$ and $B \subset C$, then $A \cup B \subset C$
(32) If $C \subset A$ and $C \subset B$, then $C \subset A \cap B$
(33) If $B \subset C$, then $A \cup B \subset A \cup C$
(34) If $B \subset C$, then $A \cap B \subset A \cap C$
(35) $A \subset A'$ if and only if $A = \Lambda$
(36) $A \cup B \neq \Lambda$ if and only if $A \neq \Lambda$ or $B \neq \Lambda$
(37) If $A \cap B \neq \Lambda$, then $A \neq \Lambda$ and $B \neq \Lambda$
(38) $A \subset B$ if and only if $B' \subset A'$
(39) $A = B$ if and only if $A \cup B \subset A \cap B$

Attentive readers will perhaps have noticed that in the foregoing discussion important use is made of the word 'object' despite its evident vagueness. Indeed, we have to a certain extent been capitalizing on that vagueness. The crux is that we are *not* using the word 'object' in such a sense that *everything* is an object, although we have no objection to letting this term refer to physical things, numbers, geometrical figures, thoughts, sensations, minds, souls, divinities, and countless other entities about the existence of which scepticism is justified. But to the question whether all sets of objects are again objects, we must refuse at this point to give a straight answer. The reason is this. We have said earlier that all the objects satisfying any sentence-form of one free variable constitute a set. Now let K be the set of just those objects satisfying

A is not an element of A.

Thus, K consists of all those objects (and only those objects) that are not elements of themselves. If K itself is an object we have the unacceptable result that K is an element of K if and only if K is not an element of K. This result is called *Russell's Antinomy* (after the English philosopher and logician, Bertrand Russell). In order to avoid it and its variants we are compelled to deviate in one way or another from what would undoubtedly be the most natural approach. From among the possible alternatives we have chosen that of giving up the assumption that all sets can themselves be members of sets, i.e., that all sets are what we are calling 'objects'.

The concept of *relation* is handled analogously to that of set; in fact, relations are treated as special kinds of sets. Any set of ordered pairs of objects is a *binary,* or two-termed relation; any set of ordered triples of objects is a *ternary* relation; any set of ordered quadruples of objects is a *quaternary* relation; in general, any set of ordered n-tuples of objects is an *n-ary* relation. A given ordered pair will belong to a binary relation R if and only if the first term of the pair is related by R to the second term; analogously for the other cases.

Let us use the notation

$$\langle x, y \rangle$$

for 'the ordered pair whose first term is x and whose second term is y', and similarly,

$$\langle x, y, z \rangle$$

for 'the ordered triple whose first term is x, second term is y, and third term is z' and so on for any finite number of occurrences of variables, names, or descriptions. Note that while

$$\{x, y\} = \{y, x\}$$

for any objects x and y, it is not true that

$$\langle x, y \rangle = \langle y, x \rangle$$

unless x and y happen to be identical with one another. Thus the ordered pair is not the same as the unordered pair or couple. It can be defined, however, as a pair of pairs, in the following way:

$$\langle x, y \rangle = \{\{x, x\}, \{x, y\}\}.$$

Thus, if we assume that all pairs of objects are themselves objects, we shall be entitled to speak of sets of ordered pairs of objects. Likewise, the ordered triple can be defined in terms of the ordered pair

$$\langle x, y, z \rangle = \langle x, \langle y, z \rangle\rangle,$$

the ordered quadruple in terms of the ordered pair and ordered triple

$$\langle x, y, z, u \rangle = \langle x, \langle y, z, u \rangle\rangle,$$

and so on. Our assumption that all pairs of objects are objects thus suffices to give us n-ary relations for every n.

Now consider a sentence-form with two free variables, e.g.,

$$x \text{ and } y \text{ are integers, and } x < y.$$

All the ordered pairs that satisfy this form will constitute a binary relation, usually called the 'less-than' relation among integers. For any objects x, y, the pair $\langle x, y \rangle$ will be a member of this relation if and only if x and y are integers and $x < y$. The pairs

$$\langle 1, 2 \rangle, \langle 2, 10 \rangle, \langle 5, 8329 \rangle, \langle 6 + 4, 11 \rangle$$

all belong to (are members of) this less-than relation, while the pairs

$$\langle 2, 1\rangle, \langle 2, 2\rangle, \langle 2^{10}, 1000\rangle, \langle \text{Laurel, Hardy}\rangle$$

do not.

Or consider the following sentence-form with three free variables:

$$x, y, z \text{ are integers, and } x < y \text{ and } y < z.$$

Corresponding to this form we have a set of ordered triples of integers, i.e., a ternary relation among integers. It consists of just those triples of integers $\langle x, y, z\rangle$ which are such that $x < y$ and $y < z$. Thus the triples

$$\langle 1, 2, 3\rangle, \langle 1, 26, 10^2\rangle$$

will belong, but the triples

$$\langle 2, 1, 3\rangle, \langle 2, 26, 26\rangle, \langle \text{Byron, Keats, Shelley}\rangle$$

will not.

A relation, being a set, is uniquely determined by its membership. Note that we *identify* the relation with the set. Hence relations that relate exactly the same objects are identical. This, no doubt, runs somewhat counter to the ordinary usage of the term 'relation'. Another point to remember is that the existence of a relation in no way depends upon whether there is a word or short phrase that denotes it. While (within appropriately specified sets) there are of course relations corresponding to 'less than', 'father of', 'brother of', 'between', 'to the left of', etc., as well as to

younger sister of mother-in-law of

and

square of the least integer that is greater than four times the positive cube root of,

we cannot expect this sort of expression to be available in every case. *Any* set of ordered pairs is a binary relation. For instance, the set consisting of the three pairs

$$\langle 1, 2\rangle, \langle 2, 1\rangle, \langle 2, 2\rangle$$

is a binary relation. So is the set consisting of the pairs

⟨Eisenhower, Rockefeller⟩, ⟨Kennedy, Johnson⟩,
⟨Goldwater, Stevenson⟩.

Likewise, any set of ordered triples is a ternary relation, whether or not it has a reasonably short name, or any name at all, in English or in any other language. In general, any set of ordered n-tuples is an n-ary relation.

In the case of binary relations a bit of special notation recommends itself: we write

$$x\,R\,y$$

instead of

$$\langle x, y \rangle \in R,$$

and similarly for other variables, names, and descriptions. Thus we would write

$$1 < 2$$

instead of

$$\langle 1, 2 \rangle \in < .$$

The *domain* of a binary relation R is the set of all objects x such that for some y, $x\,R\,y$. The *converse domain* of a binary relation R is the set of all objects y such that for some x, $x\,R\,y$. The *field* of a binary relation R is the union of its domain and its converse domain.

Since relations are sets, the unions and intersections (but not the complements) of n-ary relations are again n-ary relations. In the case of binary relations another operation is of importance. A binary relation S is called the *converse* of a binary relation R, in symbols

$$S = \breve{R},$$

if and only if, for all objects x, y,

$$x\,R\,y \text{ if and only if } y\,S\,x.$$

Thus if S is the converse of R, the couple $\langle x, y \rangle$ belongs to R if and only if the reversed couple $\langle y, x \rangle$ belongs to S.

Those binary relations that are *functions* are of special interest. A binary relation R is a function if and only if for all objects x, y, z,

$$\text{If } x\,R\,y \text{ and } x\,R\,z, \text{ then } y = z.$$

If a binary relation R and its converse are both functions, then R is called a *biunique function* or a *1–1 relation*. Relations of this sort are useful for defining sameness of number: a set A has the same number of elements as a set B if and only if there is a 1–1 relation R such that A is the domain of R and B is the converse domain of R.

The notion of an *n-ary operation* is a natural extension of the notion of a function. An $(n+1)$-ary relation R is an n-ary operation with respect to a set D if and only if, for each n-tuple $\langle x_1, x_2, \ldots, x_n \rangle$ of objects in D there is exactly one object y in D such that $\langle x_1, x_2, \ldots, x_n, y \rangle \in R$.

This completes our informal discussion of sets. In order to describe the grammar and the various possible interpretations of our artificial language we shall need to use some of the concepts which are the subject matter of set theory, and then later we may use our artificial language to characterize these concepts in a more exact way. This may seem to be a case of lifting oneself by one's own bootstraps, but perhaps it is more like using a defective machine to make a new machine that is capable of doing better work, and then using the new machine to overhaul the old one.

6. *Object-language and metalanguage.* Whenever we talk about one language by using another, we shall call the former the *object-language* (relative to that discussion) and the latter the *metalanguage*. Thus, in the case of a Greek grammar written in English, Greek is the object-language and English is the metalanguage. In the case of an English grammar written in English, English is both the object-language and the metalanguage. As these examples indicate, the distinction is relative and not absolute: what serves as metalanguage in one discussion may be the object-language in another, and the same language may at one and the same time be both. In each of the chapters to follow, an artificial language will be the object-language, and English (supplemented by certain technical terms) will be the metalanguage. Occasionally, as in Chapter 5, the artificial language and English will both be object-languages and English will be the metalanguage.

Variables of the metalanguage will be called *metalinguistic variables;* for this purpose we shall most frequently employ Greek and Fraktur letters. These metalinguistic variables are not to be confused with the variables of the artificial languages under discussion. Expressions like

$$(\phi \vee \psi)$$

are description-forms of the metalanguage. When their variables are replaced by names of expressions of the object-language, the results are descriptions of expressions of the object-language. Thus if, in the above description-form, we replace 'ϕ' by ''F^1a'' and 'ψ' by ''G^1b'', we obtain a metalinguistic expression that is an abbreviation of the following description:

> the result of writing '(', followed by 'F^1a', followed by '$\vee$', followed by 'G^1b', followed by ')',

which describes the object-language expression

$$(F^1a \vee G^1b).$$

EXERCISES

1. If quotation marks are used as suggested in section 1 of this chapter, which of the following sentences are true?
   (a) 'The Iliad' is written in English.
   (b) 'The Iliad' is an epic poem.
   (c) 'The Morning Star' and 'The Evening Star' denote the same planet.
   (d) The Morning Star is the same as the Evening Star.
   (e) '7 + 5' = '12'.
   (f) The expression ''the Campanile'' begins with a quotation mark.
   (g) The expression ''der Haifisch'' is suitable as the subject of an English sentence.

2. Using quotation marks according to our convention, punctuate the following expressions in such a way as to make them true sentences.
   (a) Saul is a name of Paul.
   (b) Mark Twain was a pseudonym of Samuel Clemens.
   (c) $2 + 2 = 4$ is a necessary truth.
   (d) $2 + 2 = 4$ is named by $2 + 2 = 4$, which in turn is named by $2 + 2 = 4$.
   (e) Although x is the 24th letter of a familiar alphabet, some authors have said x is the unknown.
   (f) The song A-sitting On a Gate is called Ways and Means although its name is The Aged Aged Man, which in turn is called Haddock's Eyes.
   (g) The quotation-name Henry is itself a quotation name of a quotation name and hence begins with two left quotes.
3. Classify each of the following as a sentence-form, a description-form, or neither.
   (a) x is wise.
   (b) 'x' is a letter of the alphabet.
   (c) the youngest brother of x.
   (d) the man who gave the diamonds to x.
   (e) x is older than the brother of y.
   (f) x is believed responsible for the loss of y.
   (g) Most people approve of x.
   (h) It is impossible that $x = 5 + 7$.
   (i) If $x < 6$, then it is impossible that $x = 5 + 7$.
   (j) the only positive integer less than 2.
   (k) the only positive integer less than the square of x.
   (l) x intends to buy y.
   (m) Ponce de Leon was hunting for x.
   (n) x prefers y to z.
   (o) x is on the grocery list.
4. How many variables occur free in each of the following sentence-forms?
   (a) $x < 6$.
   (b) $x < y$.
   (c) $x + y = x + z$, if and only if $y = z$.
   (d) There is an x such that, for every y, $x + y = z$.
   (e) $y + 1 = x$ and, for every z, $x < z$.
   (f) If for every x, $x + 1 = 1 + x$, then for some y, $y + 1 = 1 + y$.
5. For each of the following sentence-forms, name (or in some other way designate) an object that satisfies it.
   (a) x is a U. S. Senator.
   (b) $x + 1 = 4$.
   (c) x is a letter of the Greek alphabet.
   (d) x is the 24th letter of the English alphabet.
   (e) The present Queen of England has x children.
   (f) The xth time is a charm.
   (g) x wrote *Innocents Abroad.*
   (h) $x \cdot x = x$.
6. For each of the following, name a pair $\langle x, y \rangle$ that satisfies it.
   (a) x and y are cities and x is north of y.
   (b) x and y are cities and y is north of x.

(c) y is the wife of the Prime Minister of x.
(d) $x + y < x + 6$.
(e) x and y are cities and x is further from y than y is from San Francisco.

7. Let $A = \{4, 6\}$
$B = \{1, 3, 5, 7, 9\}$
$C =$ the set of positive integers that are primes
$D = \{1,\ 2 + 2,\ 2 + 1,\ 7\}$
$E = \{2, 8\}$
$F =$ the set of all objects satisfying 'x is a positive integer and $x < 10$'.
$G = \{4\}$
$H = \{1\}$
$K = \{8\}$
(a) Which of the following are true?
(1) $G \subset A$ (5) $B \subset C$
(2) $G \cup H \subset D$ (6) $A \cap D \subset G \cup H$
(3) $(A \cup B) \cup E = F$ (7) $B \subset F$
(4) $(F \cap C) \cap K = B$ (8) $A \cap B = \Lambda$
(b) Find the following:
(1) $A \cup B$ (3) $A \cap C$ (5) $A \cap (B \cup C)$ (7) $D \cap \Lambda$
(2) $A \cap B$ (4) $B \cap C$ (6) $H \cap (B \cup C)$ (8) $D \cap E$

8. Defining the difference of sets A, B (in symbols, '$A \sim B$') as follows:

$$A \sim B = A \cap B',$$

(a) which of the following hold for all sets A, B, C?
(1) $A \sim B = B \sim A$
(2) $(A \sim B) \sim C = A \sim (B \sim C)$
(3) $A \sim \Lambda = A$
(4) $A \sim V = \Lambda$
(5) $A \sim (B \cup C) = (A \sim B) \cup C$
(6) $A \sim (B \cap C) = (A \sim B) \cup (A \sim C)$
(7) $B \sim (B \sim A) = A$
(b) which of the following hold for the sets of exercise 7?
(1) $B \sim A = B \sim \Lambda$
(2) $(G \cup H) \sim H \subset B$
(3) $D \sim A \subset B$
(4) $A \sim C = A$

9. Find all the binary relations R such that the field of R is included in $\{1, 2\}$. Which of these is the 'less-than' relation among elements of $\{1, 2\}$?

In general, if a set K has n elements, how many distinct binary relations are there among the elements of K, i.e., whose fields are included in K?

10. Give an example of
(a) a sentence in which the word 'Socrates' is referred to without the use of the expression ''Socrates'';
(b) a sentence in which 'Boston' is both used and mentioned;
(c) a sentence in which numbers are mentioned;
(d) a sentence containing variables the values of which include cities;
(e) a description-form containing a variable having as substituend an expression mentioned in the form;

(f) a sentence-form not containing 'or' and satisfied by exactly two objects;
(g) a sentence-form not containing 'or' and satisfied by exactly three objects that are not numbers;
(h) two distinct sentence-forms that are satisfied by exactly the same objects.

11. Consider the following argument from the *Port Royal Logic:*

    God commands that kings must be honored.
    Louis XIV is king.
    Therefore, God commands that Louis XIV must be honored.

    Is this argument sound? Is the result of replacing the second premise by "God knows that Louis XIV is king" sound?

3

THE FORMALIZED LANGUAGE $\mathfrak{L}$

We now describe a formalized language which we shall call *the language* $\mathfrak{L}$ and with reference to which we shall study that portion of logic known technically as *elementary logic* or the *predicate calculus of first order.* Although, as will be seen later, this language is adequate for the formulation of a very wide variety of theories, its grammatical structure is in most respects far simpler than that of any natural language. In fact, the grammar of $\mathfrak{L}$ can be outlined in a couple of pages, as is done in the first section of the present chapter. In the second section we supply examples and explanatory remarks which, it is hoped, will make clear just what is intended by the several definitions presented in the first section. The third section contains some further metalinguistic terminology designed to facilitate the syntactical description of formulas and other expressions. Finally, we adopt certain notational conventions that in practice will save much writing of subscripts, superscripts, and parentheses.

1. *The grammar of* $\mathfrak{L}$. The *expressions* of the language $\mathfrak{L}$ are strings (of finite length) of symbols, which in turn are classified as follows:

A. *Variables.* The variables are the lower case italic letters 'u' through 'z', with or without numerical subscripts (i.e., subscripts that are Arabic numerals for positive integers).

B. *Constants.*

(i) The *logical constants* are the eight symbols:

$$- \quad \vee \quad (\quad) \quad \& \quad \rightarrow \quad \leftrightarrow \quad \exists$$

(ii) The *non-logical constants* fall into two classes:

(a) *Predicates,* which are capital italic letters with or without numerical subscripts and/or numerical superscripts.

(b) *Individual constants,* which are the lower-case italic letters '*a*' through '*t*', with or without numerical subscripts.

A *predicate of degree n* (or an *n-ary predicate*) is a predicate having as superscript a numeral for the positive integer n.

A *sentential letter* is a predicate without superscript.

An *individual symbol* is a variable or an individual constant.

An *atomic formula* is an expression consisting either of a sentential letter alone, or (for some positive integer n) of an n-ary predicate followed by a string (of length n) of individual symbols.

A *formula* is an expression that is either an atomic formula or else is built up from one or more atomic formulas by a finite number of applications of the following rules:

(i) If ϕ is a formula, then $-\phi$ is a formula.

(ii) If ϕ and ψ are formulas, then $(\phi \vee \psi)$, $(\phi \,\&\, \psi)$, $(\phi \rightarrow \psi)$, and $(\phi \leftrightarrow \psi)$ are formulas.

(iii) If ϕ is a formula and α is a variable, then $(\alpha)\phi$ and $(\exists\alpha)\phi$ are formulas.

Further, an occurrence of a variable α in a formula ϕ is *bound* if it is within an occurrence in ϕ of a formula of the form $(\alpha)\psi$ or of the form $(\exists\alpha)\psi$; otherwise it is a *free* occurrence.

Finally, a *sentence* is a formula in which no variable occurs free.

2. *Explanations and examples.* To the foregoing rather compressed account we hasten to add some explanations and examples. Note first of all that although we have not yet given meaning to the various symbols and formulas of $\mathfrak{L}$ (some authors prefer at this stage to speak of an 'uninterpreted calculus' instead of a 'language'), the structure of $\mathfrak{L}$ has obviously been modelled after that of the natural language. Corresponding to the words 'not', 'or', 'and', 'if ... then', and 'if and only if' of the natural language, we have in the artificial language the symbols '$-$', '$\vee$', '&', '$\rightarrow$', and '$\leftrightarrow$'. Names and other subject-expressions from the natural language, e.g., 'Socrates', are represented by the individual constants of $\mathfrak{L}$; and natural language predicates, e.g., 'is the teacher of the teacher of Aristotle', are represented by the predicates of $\mathfrak{L}$. Corresponding to sentence-forms we have the formulas of $\mathfrak{L}$. (On the other hand, note that $\mathfrak{L}$ contains no counterparts of description-forms; the more complex language $\mathfrak{L}'$, to be intro-

duced later, will differ from 𝔏 in just this respect). Formal expressions of the form (α) and $(\exists\alpha)$, where α is a variable, represent universal and existential quantifiers, respectively. And the sentences of 𝔏 are intended to be counterparts of sentences of the natural language.

We have been speaking here, in a rather vague way, of 'counterparts'. It is difficult to make this notion precise, and at the present early stage of the proceedings we shall content ourselves with a warning: do not regard the counterparts as just 'abbreviations' or 'short ways of writing' the corresponding expressions in the natural language. To be sure, we may read '&' as 'and', and 'v' as 'or', and even '→' as 'if . . . then'. But serious confusion will result if one takes these readings to indicate some sort of synonymy. In due course the whole matter of truth-conditions for the sentences of 𝔏 will be elucidated; only after that has been done will it be possible to make sense of a question whether this or that sentence of 𝔏 has (relative to a given interpretation of the symbols occurring therein) the same truth-conditions as a given sentence of the natural language.

It will turn out indeed that, whenever a pair of formal sentences ϕ and ψ have respectively the same truth-conditions as a pair of English sentences S and S' relative to a given interpretation of the formalism, then (on the whole) the formal sentence

$$(\phi \,\&\, \psi)$$

will have the same truth-conditions as the English sentence

$$S \text{ and } S'.$$

To this extent, one may say that '&' and 'and' have the same meaning. The same sort of result, but with even more reservations and qualifications, will hold of the other symbols and expressions of 𝔏, and that is what justifies the readings normally adopted. But it must be stressed that the central concepts have been defined in a purely formal way, i.e., in such a way that any sentence, formula, variable, or other expression of 𝔏 can be identified as such simply *by its shape,* without any consideration of whether or not it might 'express a complete thought', 'stand for an individual', or anything else of that sort.

The following gallery of examples will be of use.

Examples of variables of 𝔏:

$$x \quad y \quad z_1 \quad u_{26} \quad v_{398}$$

Examples of signs that are not variables of 𝔏:

$$x_0 \quad x_{\mathrm{IV}} \quad x' \quad \phi \quad \psi \quad \alpha \quad \beta \quad \Gamma \quad \Delta \quad F$$

Examples of individual constants of $\mathfrak{L}$:

$$a \quad a_1 \quad b_{16} \quad c_4 \quad t_{28}$$

Examples of signs that are not individual constants of $\mathfrak{L}$:

$$3 \quad \text{`}a\text{'} \quad \text{Scott} \quad a_0$$

Examples of predicates of $\mathfrak{L}$:

$$F \quad G_1 \quad G^2_1 \quad G^4 \quad H^{16}_{22} \quad M^{16}_{321}$$

Of the predicates just mentioned, the first two are sentential letters, the third is a predicate of degree 2, the fourth is of degree 4, and the fifth and sixth are of degree 16. All the examples given for variables and individual constants are, of course, examples of individual symbols as well.

Examples of atomic formulas of $\mathfrak{L}$:

$$F \quad G_1 \quad G^2_1ab \quad G^4x_1ya_2a_2 \quad H^1_{16}x$$

Examples of signs that are not atomic formulas of $\mathfrak{L}$:

$$x \quad y \quad Fx_1 \quad x \text{ is blue} \quad \phi \quad (F \vee G)$$

Coming now to the definition of 'formula', we note in passing that it may be stated in other words as follows:

(1) All atomic formulas are formulas.

(2) The result of prefixing the negation sign ('$-$') to any formula is again a formula.

(3) The result of inserting the wedge or the ampersand or the arrow or the double-ended arrow between two formulas (or between two occurrences of the same formula) and enclosing the whole within a pair of parentheses, is again a formula.

(4) Where α is any variable, the result of prefixing one of the expressions (α) or $(\exists\alpha)$ to any formula is again a formula.

(5) Nothing is a formula except by virtue of the foregoing four rules.

Now let us build up some examples. Since the following are atomic formulas,

$$F \quad G_1 \quad H^1x \quad G^2_1xy \quad G^2_1aa,$$

the definition informs us that they are formulas. By applying clause (i) we see that the expressions

$$-F \quad -G^2_1xy$$

are also formulas. Applying clause (ii) we may obtain such formulas as

$$(-F \rightarrow G^2_1aa) \quad (G^2_1xy \leftrightarrow G^2_1aa)$$

and, applying clause (iii), we may produce the following three formulas, among others,

$$(x)H^1x \quad (\exists x)(G_1^2xy \leftrightarrow G_1^2aa) \quad (y)(-F \rightarrow G_1^2aa).$$

Further application of clause (i) yields the two formulas

$$-(\exists x)(G_1^2xy \leftrightarrow G_1^2aa) \quad --G_1^2xy,$$

and if we continue in this way we can build up the formula

$$-(x)(\exists y)(-(\exists x)(G_1^2xy \leftrightarrow G_1^2aa) \,\&\, (-G_1^2xy \rightarrow (y)(-F \rightarrow G_1^2aa)))$$

and others of arbitrarily great complexity.

For contrast, here are a few examples of things which, appearances and misconceptions to the contrary, are not formulas of $\mathfrak{L}$.

$$Fx \quad F \vee G \quad (\phi \vee \psi) \quad F_1^2xyz \quad F_1^1x_0$$

It can be proved that the formulas of $\mathfrak{L}$ have the (very desirable) property of *unique readability,* i.e., if a formula $(\phi \rightarrow \psi)$ is the same as a formula $(\chi \rightarrow \theta)$, then ϕ and χ are the same formula, and so are ψ and θ, and similarly for conjunctions, disjunctions, biconditionals, negations, and generalizations (see section 3 for this terminology).

In order to give some examples of free and bound occurrences of variables let us consider the formula

$$(1) \qquad ((x)(F_1^2xa \rightarrow (\exists y)(F_1^2xy \,\&\, G_1^2zy)) \vee F_1^2xa).$$

In this formula there are four occurrences of the variable 'x'; the first three of these are bound, and the fourth is free. The first occurrence of 'x' in (1) is bound because it is within an occurrence in (1) of the formula

$$(2) \qquad (x)(F_1^2xa \rightarrow (\exists y)(F_1^2xy \,\&\, G_1^2zy)),$$

which is of the form $(\alpha)\psi$, where α is 'x' and ψ is the part of (2) following '(x)'. The second and third occurrences of 'x' in (1) are bound for the same reason. The fourth occurrence, however, is not bound, for it is not within an occurrence in (1) of any formula beginning with '(x)' or '$(\exists x)$'. As to the variable 'y', there are three occurrences, all bound. There is one occurrence of 'z', and that occurrence is free. For, although it is within an occurrence of (2) and also within an occurrence of

$$(3) \qquad (\exists y)(F_1^2xy \,\&\, G_1^2zy)$$

in (1), it is not within an occurrence in (1) of any formula beginning with '(z)' or '$(\exists z)$'. The individual constant 'a' occurs twice, but these occurrences are neither bound nor free. Likewise, no variable other than 'x', 'y', and 'z' occurs either bound or free in formula (1). Since, however, the formula contains at least one free occurrence of a variable, it is not a sentence.

Students are sometimes puzzled by such circumstances as the following. Although all occurrences of 'x' in '$(x)F^1x$' are bound, and 'F^1x' is a part of '$(x)F^1x$', nevertheless the occurrence of 'x' in 'F^1x' is free. The crucial consideration here is that the notions of bondage and freedom are relative and not absolute: an occurrence of a variable is bound or free *relative to a given formula;* what is bound relative to one formula may be free relative to another. Probably the confusion is further increased by the unclarity that surrounds the notion of *occurrence*. Only reluctance to introduce additional complexity prevents us from abandoning this woolly notion and defining instead a ternary relation 'α is bound at the nth place in ϕ', where 'α' takes variables as values, 'ϕ' formulas, and 'n' positive integers. Such a definition would obviate all talk about 'occurrences', but it is rather involved; consequently we hope to make do with the definition already given.

Explicit attention is called to the fact that only the expressions (α) and $(\exists\alpha)$ can bind an occurrence of the variable α; thus the occurrence of 'y' in the formula '$(x)F^1y$' is free.

3. *Additional syntactic terminology.* The syntactical description of formulas and other expressions will be greatly facilitated by introducing some further terminology. Our choice of terms is affected by the ways in which we shall later interpret the formulas of $\mathfrak{L}$, but note that the applicability of these terms, like those introduced previously, is determined solely by the shapes of the formulas involved. Again, examples follow the definitions.

For any formulas ϕ and ψ, and for any variable α,

1) $-\phi$ is called the *negation* of ϕ;
2) $(\phi \,\&\, \psi)$ is called a *conjunction,* with ϕ and ψ as *conjuncts;*
3) $(\phi \vee \psi)$ is called a *disjunction,* with ϕ and ψ as *disjuncts;*
4) $(\phi \rightarrow \psi)$ is called a *conditional,* with ϕ as *antecedent* and ψ as *consequent;*
5) $(\phi \leftrightarrow \psi)$ is called a *biconditional;*
6) $(\alpha)\phi$ is called the *universal generalization of* ϕ *with respect to* α;
7) $(\exists\alpha)\phi$ is called the *existential generalization of* ϕ *with respect to* α.

The symbols '&', 'v', '$\rightarrow$', '$\leftrightarrow$', '$-$' are called *connectives.*

A *universal quantifier* is an expression of the form (α), where α is a variable.

An *existential quantifier* is an expression of the form $(\exists\alpha)$, where α is a variable.

A formula that is not an atomic formula is called *general* if it begins with a universal or existential quantifier; otherwise it is called *molecular.*

A sentence is a *sentence of the sentential calculus* (or an *SC sentence* or an *SC formula*) if it contains no individual symbol.

An occurrence of a formula ψ in a formula ϕ is *bound* if it is properly contained within (i.e., contained within but not identical with) an occurrence in ϕ of a formula of the form $(\alpha)\theta$ or of the form $(\exists\alpha)\theta$; otherwise it is a *free* occurrence. This definition is not to be confused with the somewhat similar definition of 'bound' and 'free' as applied to variables. Note that an occurrence of (α) or $(\exists\alpha)$ may bind a formula following it even if the variable α does not occur in that formula.

In connection with later proofs it will be useful to define the *order* of a formula, as follows:

(i) Atomic formulas are of order 1.

(ii) If a formula ϕ is of order n, then $-\phi$, $(\alpha)\phi$, and $(\exists\alpha)\phi$ are of order $n + 1$, where α is a variable.

(iii) If the maximum of the orders of formulas ϕ and ψ is n, then $(\phi \mathbin{\&} \psi)$, $(\phi \vee \psi)$, $(\phi \rightarrow \psi)$, and $(\phi \leftrightarrow \psi)$ are of order $n + 1$.

Finally, for any formula ϕ, variable α, and individual symbol β, $\phi\ \alpha/\beta$ is the result of replacing all free occurrences of α in ϕ by occurrences of β.

To illustrate the use of the foregoing terminology we can employ some of it in a description of the long formula (1), which was considered in connection with free and bound occurrences of variables. That formula is a disjunction, one of the disjuncts of which is the atomic formula 'F^2_1xa', while the other disjunct is the universal generalization with respect to 'x' of a conditional having as antecedent the formula 'F^2_1xa' and having as consequent the existential generalization with respect to 'y' of a conjunction, of which one of the conjuncts is 'F^2_1xy', while the other is 'G^2_1zy'. In that same formula the first occurrence of 'F^2_1xa' is a bound occurrence (and note that it would still be bound if the preceding quantifier were '(y)' instead of '(x)'), and the second occurrence is free. The occurrence of

$$(\exists y)(F^2_1xy \mathbin{\&} G^2_1zy)$$

is bound, but the occurrence of

$$(x)(F^2_1xa \rightarrow (\exists y)(F^2_1xy \mathbin{\&} G^2_1zy))$$

is free, as is the occurrence of the whole formula. The phrase 'properly contained within' in this definition means 'contained within and not identical with'; if we speak of the formulas that are contained within a given formula, we intend to include, in particular, the given formula itself; but if we speak of the formulas that are *properly* contained within a given formula, we intend to exclude the given formula itself.

Examples of atomic sentences:

$$F^1a \quad G^4_2abab \quad H^6_1abc_2c_2ba$$

Examples of general sentences:

$$(x)F^1x \quad (x)(\exists y)(F^2_1xy \,\&\, G^3_2yxb)$$

Examples of molecular sentences:

$$(F \vee G) \quad ((x)F^1x \,\&\, (\exists y)G^1y) \quad --P$$

The five formulas on the two displayed lines above are of orders 2, 4, 2, 3, and 3, respectively.

Finally, if ϕ is 'F^1x', α is 'x', and β is 'a', then $\phi\ \alpha/\beta$ is 'F^1a'. Again, if ϕ is 'F^2xa', α is 'x', and β is 'a', then $\phi\ \alpha/\beta$ is 'F^2aa'. If ϕ is '$(F^1x \vee (x)F^1x)$', α is 'x', and β is 'b', then $\phi\ \alpha/\beta$ is '$(F^1b \vee (x)F^1x)$'. If ϕ is '$(x)F^1x$', α is 'x', and β is 'a', then $\phi\ \alpha/\beta$ is '$(x)F^1x$'. And if ϕ is 'F^1a', α is 'x', β is 'y', then $\phi\ \alpha/\beta$ is 'F^1a'.

4. *Notational conventions.* In writing formulas we shall occasionally omit the outermost parentheses, 'but only when no confusion threatens to result', as the saying goes. In addition, and under the same proviso, it will be convenient informally to omit superscripts and/or subscripts from predicates. For example, we may write

$$(x)Fx \rightarrow Fa$$

instead of

$$((x)F^1x \rightarrow F^1a),$$

though *not* instead of

$$((x)F^1_1x \rightarrow F^1_2a).$$

It must be remembered at all times, however, that our formal rules and definitions apply only to formulas written out in full. Thus, although the first expression just quoted begins with the quantifier '(x)', the sentence that it abbreviates is not a general sentence, for *it* does *not* begin with a quantifier.

EXERCISES

1. Using the definition of 'formula' literally (i.e., making no use of the conventions about dropping parentheses, superscripts, and subscripts) state in each of the following cases (a) whether the given expression is a formula, (b) whether it is a sentence, (c) whether it is an atomic sentence, a general sentence, a molecular sentence, or none of these, (d) whether 'x' occurs free in it, (e) whether 'x' occurs bound in it. Tabulate your answers.
   (1) F^1x
   (2) $(x)F^1x$

(3) F^1a
(4) $(x)F^1a$
(5) $-(x)F^1x \vee F^1a$
(6) $((x)F_1^1x \rightarrow F_1^1a)$
(7) $(x)(y)(F^1xy \rightarrow F^1yx)$
(8) $((\exists x)(y)G_1^2xy \rightarrow (y)(\exists x)G_1^2xy)$
(9) $((x)F^1x \vee -F^1x)$
(10) $(x)(F_1^1x \vee -F_1^1x)$
(11) $(F^1a \rightarrow (-F^1a \rightarrow F^1a))$
(12) $(P \rightarrow (x)P)$
(13) $(((F^1a \leftrightarrow F^1x) \leftrightarrow F^1a) \leftrightarrow F^1x)$
(14) $(-F^1a) \& (-F^1b)$

2. Letting α be 'x' and β be 'b', write $\phi\ \alpha/\beta$ when ϕ is formula (1) above. Do the same for $\phi = (6)$, $\phi = (9)$, $\phi = (12)$, $\phi = (13)$.
3. Give an example of:
   (a) a formula in which 'x' and 'y' occur free;
   (b) a sentence that begins with a universal quantifier;
   (c) a sentence that is a conjunction, the conjuncts of which are disjunctions, the disjuncts of which are sentential letters or negations of sentential letters;
   (d) a formula in which no individual symbol occurs;
   (e) a formula that is a universal generalization of an existential generalization of a conditional;
   (f) a formula that is not a sentence but has a part that is a sentence;
   (g) a formula that is a sentence but has no other part that is a sentence.
4. Insert quotation marks so as to make the following true:

   Although α is not a variable of $\mathfrak{L}$ and β is not an individual symbol of $\mathfrak{L}$, there are α, β, and ϕ such that α is a variable of $\mathfrak{L}$, β is an individual constant of $\mathfrak{L}$, ϕ is a formula of $\mathfrak{L}$, and $\phi\ \alpha/\beta$ is the same as ϕ.
5. For each of the following conditions, give a pair of distinct (i.e., non-identical) formulas ϕ and ψ that satisfy it.
   (a) ϕ is the result of replacing all free occurrences of 'x' in ψ by occurrences of 'a', and ψ is the result of replacing all occurrences of 'a' in ϕ by occurrences of 'x'.
   (b) ϕ is the result of replacing all free occurrences of 'x' in ψ by occurrences of 'a', but ψ is not the result of replacing all occurrences of 'a' in ϕ by occurrences of 'x'.
   (c) ϕ is like ψ except that ϕ contains free occurrences of 'y' where, but not only where, ψ contains free occurrences of 'x'.
6. Show that every formula ϕ contains exactly as many occurrences of '(' as occurrences of ')'. Use the (strong) principle of mathematical induction (see page 168); i.e., show that, for any positive integer k, if the assertion holds for formulas of order $<k$, then it holds for formulas of order k.
7. Show, again by mathematical induction on the order of formulas, that no proper initial segment of a formula (i.e., an initial segment that is not identical with the whole formula) is again a formula. You may use the result of exercise 6. Does the same hold for proper final segments?

8. Show by mathematical induction that the following two conditions are equivalent for all formulas ϕ, ψ.
   (a) ϕ is like ψ except for containing free occurrences of 'y' where and only where ψ contains free occurrences of 'x'.
   (b) ϕ is like ψ except for containing free occurrences of 'y' wherever ψ contains free occurrences of 'x', and ψ contains no free occurrences of 'y'.

4

INTERPRETATIONS AND VALIDITY

In our choice of symbols and readings, and in our use of the words 'predicate', 'sentence', etc., we have been looking toward the possibility of assigning meanings to the expressions of $\mathfrak{L}$. The present chapter is devoted to such assignments (or 'interpretations', as we shall call them) and to certain important concepts that may be defined with their help. The most important of these are the concepts of consequence and validity. A sentence of $\mathfrak{L}$ is valid if it comes out true no matter what meanings (more precisely, denotations) are assigned to its non-logical constants—in other words, if its truth depends only upon the semantic properties of its logical framework. And one sentence is a consequence of another just in case (i.e., if and only if) the corresponding conditional is valid.

The order in which these matters are presented is as follows. First we define the notion of an interpretation of the language $\mathfrak{L}$. Then we explain just what is meant by saying that a sentence of $\mathfrak{L}$ is true relative to a given interpretation. The valid sentences are then defined as the sentences true under all interpretations, and related definitions are given for 'consequence' and 'consistency'. Finally we list some of the most important properties of the concepts thus defined.

1. *Interpretations of* $\mathfrak{L}$. We shall first characterize the situation in a rough way and then attempt to be more precise. Given a sentence ϕ of $\mathfrak{L}$, an in-

terpretation assigns a denotation to each non-logical constant occurring in ϕ. To individual constants it assigns individuals (from some universe of discourse); to predicates of degree 1 it assigns properties (more precisely, sets) of individuals; to predicates of degree 2 it assigns binary relations of individuals; to predicates of degree 3 it assigns ternary relations of individuals, and so on; and to sentential letters it assigns truth-values. Under such an interpretation atomic sentences are considered as stating that the individuals assigned to their individual constants stand in the relations assigned to their predicates (or, if the atomic sentence is a sentential letter, it is considered as standing for the truth-value assigned to it). When in addition the logical constants occurring in ϕ are understood in the way usual among logicians (i.e., '$\vee$' as standing for 'or', '$-$' for 'not', the universal quantifier for 'all', etc.), we find that ϕ is true or false and hence, to that extent at least, is meaningful.

A few examples will indicate what is here intended. Consider the sentence 'D^1s'. If, according to a particular interpretation $\mathfrak{I}$ that has the set of human beings as its universe of discourse, the individual constant 's' denotes Socrates and the predicate 'D^1' denotes the set of all persons who died in 399 B.C., then under that interpretation the sentence 'D^1s' states that Socrates belongs to the set of all persons who died in 399 B.C., i.e., that Socrates died in 399 B.C., and accordingly it is true. Further, the sentence '$-D^1s$' states under this interpretation that Socrates did not die in 399 B.C., and so it is false. Under the same interpretation the sentence '$(D^1s \vee -D^1s)$' says that either Socrates died in 399 B.C. or he did not die in 399 B.C.; hence, it again is true. Likewise, under the interpretation $\mathfrak{I}$ the sentence '$(\exists x)D^1x$' says that somebody died in 399 B.C., which is true; while '$(x)D^1x$' says that everyone died in 399 B.C., which is false.

Consider next a different interpretation $\mathfrak{I}'$, which assigns to 's' the individual Sir Walter Scott and to 'D^1' the set of all persons who wrote *The Daffodils*. Under this interpretation 'D^1s' is false; '$-D^1s$' is true; '$(D^1s \vee -D^1s)$' is true; '$(\exists x)D^1x$' is true; '$(x)D^1x$' is false. Thus we see that the truth-value of a sentence may change as we go from one interpretation to another, and in most cases it certainly will.

There are sentences, however, for which the truth-value will not change from one interpretation to another. For instance, the sentence '$(D^1s \vee -D^1s)$', which we have found to be true both under $\mathfrak{I}$ and under $\mathfrak{I}'$, will be true under every interpretation. For no matter what individual is assigned to 's' and no matter what set is assigned to 'D^1', the given sentence will say that the individual either is in the set or is not in the set, and hence it will be true. Sentences that are true under every interpretation are called 'valid' sentences of $\mathfrak{L}$. They are the exception rather than the rule, of course, but they are of crucial importance to logic.

We proceed now to a more careful presentation of these matters. It proves convenient to define 'interpretation' in such a way that each interpretation assigns denotations to *all* the non-logical constants of the language $\mathfrak{L}$, and not just to a few that happen to occur in sentences under consideration at the moment. We shall say that to give an interpretation of our artificial language is (1) to specify a non-empty domain $\mathfrak{D}$ (i.e., a non-empty set) as the universe of discourse; (2) to assign to each individual constant an element of $\mathfrak{D}$; (3) to assign to each n-ary predicate an n-ary relation among elements of $\mathfrak{D}$; and (4) to assign to each sentential letter one of the truth-values T (truth) or F (falsehood).

Thus, an *interpretation* of $\mathfrak{L}$ consists of a non-empty domain $\mathfrak{D}$ together with an assignment that associates with each individual constant of $\mathfrak{L}$ an element of $\mathfrak{D}$, with each n-ary predicate of $\mathfrak{L}$ an n-ary relation among elements of $\mathfrak{D}$, and with each sentential letter of $\mathfrak{L}$ one of the truth-values T or F. (In this definition we consider a singulary relation among elements of $\mathfrak{D}$ to be simply a set of elements of $\mathfrak{D}$, a binary relation among elements of $\mathfrak{D}$ to be a set of ordered pairs of elements of $\mathfrak{D}$, a ternary relation among elements of $\mathfrak{D}$ to be a set of ordered triples, and so on.)

In the definitions of 'valid' and 'true', which will be given later, there is reference to *all* interpretations of a certain type. To understand fully the effect of this reference, the student must bear in mind that *any* non-empty set may be chosen as the domain of an interpretation, and that *all* n-ary relations among elements of the domain are candidates for assignment to any predicate of degree n. It is not necessary that the domain be a set with a natural-sounding name, like the set of human beings or the set of even integers; it could be the set $\{0, 4, 327\}$ or the set consisting exclusively of William Shakespeare and Albert Einstein. Nor do the sets and relations assigned to predicates need to have natural-sounding names, like 'less than' or 'father of'. If the domain were $\{0, 4, 327\}$, for instance, we could assign to the binary predicate 'F^2' the binary relation

$$\{\langle 0, 4\rangle, \langle 4, 0\rangle, \langle 327, 327\rangle\},$$

or the binary relation

$$\{\langle 0, 4\rangle\},$$

or the binary relation Λ, or the binary relation consisting of all ordered pairs that can be formed from elements of the domain. Note also that an interpretation may assign the same relation simultaneously to more than one predicate (though never, of course, different relations to the same predicate). Similar remarks hold of the individual constants and sentential letters. These matters are mentioned explicitly here because there is a natural tendency among beginners to consider only those sets and relations for which there exist compact and usual expressions in everyday language.

2. *Truth.* Next we wish to make clear what it means to say that a sentence of $\mathfrak{L}$ is true or false with respect to a given interpretation of $\mathfrak{L}$. In other words, we seek to give an exact definition of the locution

$$\phi \text{ is true under } \mathfrak{I},$$

where the values of 'ϕ' are sentences of $\mathfrak{L}$ and the values of '$\mathfrak{I}$' are interpretations of $\mathfrak{L}$.

The complete definition is rather long, and the clauses dealing with quantifiers involve certain complications. Before presenting it *in toto* we shall consider how it *would* run if we could restrict ourselves to sentences containing no quantifiers.

As is obvious from the definitions of 'sentence' and 'formula', any sentence of $\mathfrak{L}$ that contains no quantifiers is either an atomic sentence or is built up out of atomic sentences by means of the connectives. (By contrast, sentences that do contain quantifiers may lack this property; for example, the sentence '$(\exists x)(Fx \mathbin{\&} Gx)$' is not atomic and yet does not contain any atomic sentence as a part). Thus if, relative to a given interpretation $\mathfrak{I}$, we state the conditions under which atomic sentences are true, and if we then indicate how the truth-values of molecular sentences are determined by the truth-values of their parts, we shall have defined truth with respect to $\mathfrak{I}$ for all sentences in which no quantifiers occur.

Such a definition would go as follows. Let $\mathfrak{I}$ be any interpretation and ϕ be any quantifier-free sentence of $\mathfrak{L}$. Then

1) if ϕ is a sentential letter, then ϕ is true under $\mathfrak{I}$ if and only if $\mathfrak{I}$ assigns T to ϕ; and

2) if ϕ is atomic and is not a sentential letter, then ϕ is true under $\mathfrak{I}$ if and only if the objects that $\mathfrak{I}$ assigns to the individual constants of ϕ are related (when taken in the order in which their corresponding constants occur in ϕ) by the relation that $\mathfrak{I}$ assigns to the predicate of ϕ; and

3) if $\phi = -\psi$, then ϕ is true under $\mathfrak{I}$ if and only if ψ is not true under $\mathfrak{I}$; and

4) if $\phi = (\psi \vee \chi)$ for sentences ψ, χ, then ϕ is true under $\mathfrak{I}$ if and only if ψ is true under $\mathfrak{I}$ or χ is true under $\mathfrak{I}$ or both; and

5) if $\phi = (\psi \mathbin{\&} \chi)$ for sentences ψ, χ, then ϕ is true under $\mathfrak{I}$ if and only if ψ is true under $\mathfrak{I}$ and χ is true under $\mathfrak{I}$; and

6) if $\phi = (\psi \rightarrow \chi)$ for sentences ψ, χ, then ϕ is true under $\mathfrak{I}$ if and only if either ψ is not true under $\mathfrak{I}$, or χ is true under $\mathfrak{I}$, or both; and

7) if $\phi = (\psi \leftrightarrow \chi)$ for sentences ψ, χ, then ϕ is true under $\mathfrak{I}$ if and only if either ψ and χ are both true or both not true under $\mathfrak{I}$.

Further, ϕ is *false* under $\mathfrak{I}$ if and only if ϕ is not true under $\mathfrak{I}$.

Before looking at definite examples, let us paraphrase the definition so as to minimize such difficulties as may be caused by the notation.

According to clause 1, an atomic sentence that is just a sentential letter is to count as true under an interpretation if and only if the interpretation assigns the value T to that letter. For this case the definition may seem trivial. Clause 2 applies to the remaining atomic sentences, each of which will consist of an *n*-ary predicate followed by *n* occurrences of individual constants. (Obviously, no atomic sentence can contain a variable). It states that an atomic sentence of this form will be true under a given interpretation if the objects denoted by the individual constants stand in the relation denoted by the predicate. Clause 3 adds that the negation of a sentence is true under an interpretation if and only if the sentence is not true (i.e., false) under that interpretation; 4 states that a disjunction of two sentences is true if and only if one or both of the disjuncts is true; 5, that a conjunction is true if and only if both conjuncts are true; 6, that a conditional is true if and only if the antecedent is false or the consequent true, or both, (i.e., unless the antecedent is true and the consequent false, in which case the conditional is false); and 7, that a biconditional is true if and only if both parts have the same truth-value.

It is important to see that since every quantifier-free sentence is an atomic sentence or is a molecular compound of shorter sentences, it will fall under one of the seven clauses. If it falls under clauses 1 or 2, its truth relative to a given interpretation is dealt with explicitly, whereas if it falls under one of the clauses 3 to 7 its truth relative to a given interpretation is reduced to the truth or falsity of simpler sentences under that interpretation. These simpler sentences will in turn fall under clauses 1 to 7, so that eventually, since no sentence can contain more than a finite number of occurrences of connectives, the question of the truth or falsity of the original sentence under a given interpretation will be reduced to questions as to the truth or falsity of atomic sentences under that interpretation; and these latter questions are settled explicitly in clauses 1 and 2.

So much for paraphrasing and general explanation. Let us now see how the definition applies to particular examples. In order to do this we must give some interpretations of $\mathfrak{L}$ and consider some sentences. Accordingly, let $\mathfrak{I}$ be an interpretation having the set of all positive integers as its domain and assigning denotations to non-logical constants as follows:

E^1: the set of even positive integers
O^1: the set of odd positive integers
P^1: the set of prime positive integers
L^2: the binary relation that holds between positive integers $\mathfrak{m}$, $\mathfrak{n}$ when $\mathfrak{m} < \mathfrak{n}$; i.e., the relation 'less-than'
I^2: the binary relation of identity between positive integers
S^3: the ternary relation that holds among positive integers $\mathfrak{m}$, $\mathfrak{n}$, $\mathfrak{p}$ when $\mathfrak{m} + \mathfrak{n} = \mathfrak{p}$
M^3: the ternary relation that holds among positive integers $\mathfrak{m}$, $\mathfrak{n}$, $\mathfrak{p}$ when $\mathfrak{m} \cdot \mathfrak{n} = \mathfrak{p}$

a_1: 1 P: T
a_2: 2 Q: F
a_3: 3 R: T
. .
. .
. .

(Note that the foregoing does not strictly 'give' an interpretation of $\mathfrak{L}$, since the assignments to most of the non-logical constants of $\mathfrak{L}$ are left unspecified. But, as will be seen later, it turns out that the truth or falsity of sentences under a given interpretation depends only upon what that interpretation assigns to the constants actually occurring in those sentences.)

Now take the sentence

(1) $$La_1a_2.$$

This is an atomic sentence; it is not a sentential letter, and thus clause 2 is the applicable clause. Applying 2 we find that 'La_1a_2' is true under $\mathfrak{I}$ if and only if the objects that $\mathfrak{I}$ assigns to 'a_1' and 'a_2' (i.e., the positive integers 1 and 2) are related by the relation that $\mathfrak{I}$ assigns to 'L' (i.e., the relation 'less-than'). In other words, 'La_1a_2' is true under $\mathfrak{I}$ if and only if $1 < 2$. Similarly, 'La_2a_1' is true under $\mathfrak{I}$ if and only if $2 < 1$, and 'La_1a_1' is true under $\mathfrak{I}$ if and only if $1 < 1$.

Consider next the sentence

(2) $$La_1a_2 \vee La_2a_1.$$

This sentence is a disjunction; clause 4 applies. It tells us that '$La_1a_2 \vee La_2a_1$' is true under $\mathfrak{I}$ if and only if 'La_1a_2' is true under $\mathfrak{I}$ or 'La_2a_1' is true under $\mathfrak{I}$, or both. Making use of the conditions just obtained for these atomic sentences, we have: '$La_1a_2 \vee La_2a_1$' is true under $\mathfrak{I}$ if and only if $1 < 2$ or $2 < 1$ or both.

In the same way, applying clauses 6, 3, and 2, we find that

(3) $$La_1a_2 \rightarrow -La_2a_1$$

is true under $\mathfrak{I}$ if and only if $1 \nless 2$ or $2 \nless 1$. (Provided that 'if ... then' is understood in the so-called 'material' sense, we can state this same condition as 'if $1 < 2$, then $2 \nless 1$'.)

For the sentence

(4) $$(Sa_2a_2a_5 \; \& \; Sa_2a_2a_4) \rightarrow Ia_4a_5,$$

applying clauses 6, 5, and 2, we have: (4) is true under $\mathfrak{I}$ if and only if, if $2 + 2 = 5$ and $2 + 2 = 4$, then $4 = 5$.

It is time now to return to the basic task of this section, which is to define 'true under $\mathfrak{I}$' for all sentences of $\mathfrak{L}$, including those that contain quantifiers. To deal with the quantifiers there is need of an auxiliary notion defined as follows:

Let $\mathfrak{I}$ and $\mathfrak{I}'$ be interpretations of $\mathfrak{L}$, and let β be an individual constant; then $\mathfrak{I}$ is a *β-variant* of $\mathfrak{I}'$ if and only if $\mathfrak{I}$ and $\mathfrak{I}'$ are the same or differ only in what they assign to β. (This implies, be it noted, that if $\mathfrak{I}$ is a β-variant of $\mathfrak{I}'$, then $\mathfrak{I}$ and $\mathfrak{I}'$ have the same domain.)

The definition of the locution *true under* $\mathfrak{I}$ may now be formulated. As in the restricted case it proceeds recursively: it states explicitly the conditions under which the simplest sentences are true, and then it indicates how the truth-values of complex sentences depend upon those of simpler ones (which this time are not necessarily their parts). Accordingly, let ϕ be any sentence of $\mathfrak{L}$, α a variable, and β the first* individual constant not occurring in ϕ. Then

1) if ϕ is a sentential letter, then ϕ is true under $\mathfrak{I}$ if and only if $\mathfrak{I}$ assigns T to ϕ; and

2) if ϕ is atomic and is not a sentential letter, then ϕ is true under $\mathfrak{I}$ if and only if the objects that $\mathfrak{I}$ assigns to the individual constants of ϕ are related (when taken in the order in which their corresponding constants occur in ϕ) by the relation that $\mathfrak{I}$ assigns to the predicate of ϕ; and

3) if $\phi = -\psi$, then ϕ is true under $\mathfrak{I}$ if and only if ψ is not true under $\mathfrak{I}$; and

4) if $\phi = (\psi \vee \chi)$ for sentences ψ, χ, then ϕ is true under $\mathfrak{I}$ if and only if ψ is true under $\mathfrak{I}$ or χ is true under $\mathfrak{I}$ or both; and

5) if $\phi = (\psi \,\&\, \chi)$ for sentences ψ, χ, then ϕ is true under $\mathfrak{I}$ if and only if ψ is true under $\mathfrak{I}$ and χ is true under $\mathfrak{I}$; and

6) if $\phi = (\psi \rightarrow \chi)$ for sentences ψ, χ, then ϕ is true under $\mathfrak{I}$ if and only if either ψ is not true under $\mathfrak{I}$, or χ is true under $\mathfrak{I}$, or both; and

7) if $\phi = (\psi \leftrightarrow \chi)$ for sentences ψ, χ, then ϕ is true under $\mathfrak{I}$ if and only if either ψ and χ are both true or both not true under $\mathfrak{I}$; and

8) if $\phi = (\alpha)\psi$, then ϕ is true under $\mathfrak{I}$ if and only if $\psi\ \alpha/\beta$ is true under every β-variant of $\mathfrak{I}$; and

9) if $\phi = (\exists\alpha)\psi$, then ϕ is true under $\mathfrak{I}$ if and only if $\psi\ \alpha/\beta$ is true under at least one β-variant of $\mathfrak{I}$.

Further, ϕ is *false under* $\mathfrak{I}$ if and only if ϕ is not true under $\mathfrak{I}$.

(Instead of saying that ϕ is true under $\mathfrak{I}$ or that ϕ is false under $\mathfrak{I}$ we shall sometimes say that $\mathfrak{I}$ *assigns the truth-value T to* ϕ or that $\mathfrak{I}$ *assigns the truth-value F to* ϕ. Still other ways of saying that ϕ is true under $\mathfrak{I}$ are to say that $\mathfrak{I}$ *is a model of* ϕ or that $\mathfrak{I}$ *satisfies* ϕ. This terminology is also extended to sets of sentences Γ; e.g., to say that $\mathfrak{I}$ satisfies Γ is to say that every sentence in Γ is true under $\mathfrak{I}$.)

Again, we hasten to offer explanations and examples. The only difference between the definition just given and the restricted one previously considered is in the addition of clauses 8 and 9, which apply to sentences

*We suppose that the individual constants are listed in the following order: $a, b, \ldots, t, a_1, b_1, \ldots, t_1, a_2, b_2, \ldots$.

that are general. Any such sentence will consist of a quantifier, (α) or $(\exists\alpha)$, followed by a formula ψ in which no variable other than α occurs free. (If some variable other than α did occur free in the formula ψ, then neither $(\alpha)\psi$ nor $(\exists\alpha)\psi$ would be a sentence, since, with the sole exception of α, any free variable in ψ remains free after (α) or $(\exists\alpha)$ has been prefixed.) Thus if we go through ψ replacing all free occurrences of α by occurrences of some individual constant β—and for definiteness we take as β the first individual constant not occurring already in ψ—the result will be a sentence ψ' (i.e., $\psi\ \alpha/\beta$). Now the sentence ψ', as contrasted with the formula ψ, will have a truth-value under any interpretation one chooses. Moreover, if, as is natural, we take the number of occurrences of connectives and quantifiers as a measure of the complexity of a formula, ψ' is simpler than $(\alpha)\psi$ and $(\exists\alpha)\psi$. Thus something will be gained by reducing questions of the truth or falsity of $(\alpha)\psi$ and $(\exists\alpha)\psi$ to questions of the truth or falsity of ψ'. This is just what clauses 8 and 9 do. According to 8, $(\alpha)\psi$ is true under an interpretation $\mathfrak{I}$ just in case ψ' is true not only under $\mathfrak{I}$ itself but also under all other β-variants of $\mathfrak{I}$; analogously, clause 9 says that $(\exists\alpha)\psi$ is true under $\mathfrak{I}$ if and only if ψ' is true under $\mathfrak{I}$ or under some other β-variant of $\mathfrak{I}$.

Consider how these clauses operate when applied to concrete cases. Let $\mathfrak{I}$ be the interpretation given in connection with our discussion of the restricted definition, and consider the sentence

$$(5) \qquad (\exists y)La_2y.$$

By clause 9 we ascertain that (5) is true under $\mathfrak{I}$ if and only if 'La_2a' (which is the result of replacing all free occurrences of 'y' in 'La_2y' by occurrences of the first individual constant not occurring in 'La_2y', viz., 'a') is true under some 'a'-variant of $\mathfrak{I}$. Now the 'a'-variants of $\mathfrak{I}$, according to the definition, are $\mathfrak{I}$ itself together with all interpretations that differ from $\mathfrak{I}$ only in what they assign to the individual constant 'a'. Thus one 'a'-variant of $\mathfrak{I}$ will be just like $\mathfrak{I}$ except for assigning the integer 2 to 'a' (as well as to 'a_2', of course); another will be like $\mathfrak{I}$ except for assigning 3 to 'a'; and so on. No matter what integer $\mathfrak{n}$ is chosen, there will be some 'a'-variant of $\mathfrak{I}$ assigning that integer to 'a', and, furthermore, 'La_2a' will be true under that 'a'-variant if and only if $2 < \mathfrak{n}$. Thus suppose $\mathfrak{I}_2$ is the 'a'-variant of $\mathfrak{I}$ that assigns 2 to 'a'. Then 'La_2a' will be true under $\mathfrak{I}_2$ if and only if $2 < 2$. Let $\mathfrak{I}_6$ be the 'a'-variant of $\mathfrak{I}$ that assigns 6 to 'a'. Then 'La_2a' will be true under $\mathfrak{I}_6$ if and only if $2 < 6$. In general, for any positive integer $\mathfrak{n}$, if $\mathfrak{I}_\mathfrak{n}$ is the 'a'-variant of $\mathfrak{I}$ that assigns $\mathfrak{n}$ to 'a', then 'La_2a' is true under $\mathfrak{I}_\mathfrak{n}$ if and only if $2 < \mathfrak{n}$.

From this we see that 'La_2a' is true under some 'a'-variant of $\mathfrak{I}$ if and only if there is some positive integer $\mathfrak{n}$ such that $2 < \mathfrak{n}$; in other words, (5) is true under $\mathfrak{I}$ if and only if there is a positive integer $\mathfrak{n}$ such that $2 < \mathfrak{n}$.

This, of course, is in accord with our intended understanding of the existential quantifier.

In connection with clause 8, consider

(6) $$(x)(\exists y)Lxy.$$

This is of the form $(\alpha)\psi$, where $\alpha =$ 'x' and $\psi =$ '$(\exists y)Lxy$'; therefore, clause 8 applies. We are there informed that (6) is true under $\mathfrak{I}$ if and only if '$(\exists y)Lay$' is true under every 'a'-variant of $\mathfrak{I}$. By considerations analogous to those presented in the case of (5), it is easy to establish that '$(\exists y)Lay$' is true under the 'a'-variant $\mathfrak{I}_{\mathfrak{m}}$ of $\mathfrak{I}$ just in case the integer $\mathfrak{m}$ is less than some integer. Therefore, '$(\exists y)Lay$' will be true under *all* 'a'-variants of $\mathfrak{I}$ if and only if, for *every* integer $\mathfrak{m}$, there is an integer $\mathfrak{n}$ such that $\mathfrak{m} < \mathfrak{n}$. We consequently have the result, again in harmony with our antecedent intentions, that (6) is true under $\mathfrak{I}$ if and only if, for every positive integer $\mathfrak{m}$, there is a positive integer $\mathfrak{n}$ such that $\mathfrak{m} < \mathfrak{n}$.

It is hoped, of course, that the foregoing discussion will help the student to understand our definition of 'true under $\mathfrak{I}$'. It is also hoped, and even expected, that he will soon 'get the idea' of the definition and will then no longer need to apply it in the grinding way illustrated above. Indeed, when the sentences are complex, its step-by-step application becomes a task for which life is too short.

Actually, the definition finds its principal justification as an exact point of reference on the basis of which one can establish certain general and fundamental principles that are frequently applied; among these are the following (we temporarily restrict ourselves to sentences and abbreviate 'true under $\mathfrak{I}$' and 'false under $\mathfrak{I}$' as 'true' and 'false'):

1) A negation is true if and only if what is negated is false.

2) A disjunction is true if and only if at least one of the disjuncts is true.

3) A conjunction is true if and only if both of the conjuncts are true.

4) A conditional is true if and only if either the antecedent is false or the consequent true, or both.

5) A biconditional is true if and only if both parts have the same truth-value.

6) If a universal generalization $(\alpha)\phi$ is true, then, for every constant β, $\phi\ \alpha/\beta$ is true. (The converse of this does not hold.)

7) If, for some constant β, $\phi\ \alpha/\beta$ is true, then $(\exists\alpha)\phi$ is true. (Again, the converse does not hold.)

It is perhaps worth mentioning in passing that there is a way of defining the truth of general sentences without getting into complexities like those associated with the notion of β-variant. The idea would be to define

$$(x)Fx,$$

for example, as true if and only if all the sentences

$$Fa,\ Fb,\ \ldots,\ Ft,\ Fa_1,\ Fb_1,\ \ldots,\ Ft_1,\ Fa_2,\ \ldots,\ \ldots$$

are true. Perhaps a brief consideration of this alternative will throw further light on the method we have used.

Note first of all that from an intuitive point of view the new approach is plausible only if each object in the universe of discourse is named by at least one of the constants in the list

$$a, b, \ldots, t, a_1, b_1, \ldots, t_1, a_2, b_2, \ldots, t_2, a_3, \ldots.$$

Otherwise it could happen that all the 'F-sentences' mentioned above were true, but that nevertheless some object in the domain was not in the set associated with 'F', in which case we would still not wish to say that

$$(x)Fx$$

was true. Bearing this in mind, let us single out from the totality of interpretations those interpretations $\mathfrak{I}$ which are such that, for each element of the domain of $\mathfrak{I}$, there is an individual constant to which $\mathfrak{I}$ assigns that element as denotation (in other words, every element of the domain of $\mathfrak{I}$ is assigned a name by $\mathfrak{I}$); and let us call these *complete interpretations*. Then clauses 8 and 9 in the definition of 'true under $\mathfrak{I}$' may be replaced by

8′) if $\phi = (\alpha)\psi$, then ϕ is true under $\mathfrak{I}$ if and only if every sentence $\psi\ \alpha/\beta$, where β is an individual constant, is true under $\mathfrak{I}$; and

9′) if $\phi = (\exists\alpha)\psi$, then ϕ is true under $\mathfrak{I}$ if and only if some sentence $\psi\ \alpha/\beta$, where β is an individual constant, is true under $\mathfrak{I}$.

It can be shown that for any sentence ϕ, ϕ is true under all complete interpretations if and only if ϕ is true under all interpretations (cp. the definition of 'valid' below).* Nevertheless, since in applying $\mathfrak{L}$ we do not wish to be restricted to domains for which there are in $\mathfrak{L}$ enough individual constants to go around, we shall stick to a definition which does not presuppose that every element in the universe of discourse has a name. We may wish, for instance, to formulate in $\mathfrak{L}$ the theory of real numbers, even though it is known, by a famous proof of Cantor, that there is no way in which the set of real numbers can be put into 1-1 correspondence with the set of individual constants of $\mathfrak{L}$.

3. *Validity, consequence, consistency*. In terms of the notion of 'true under an interpretation' it is possible to define three of the most important concepts in logic.

A sentence ϕ is *valid* (or *logically true*) if and only if ϕ is true under every interpretation.

A sentence ϕ is a *consequence* of a set of sentences Γ if and only if there

* But, as was pointed out by Belnap and Dunn (*Nous* 2 (1968), pp. 177ff.), it is not the case that, for any sentence ϕ and set of sentences Γ, ϕ is true under all complete interpretations under which Γ is true if and only if ϕ is true under all interpretations under which Γ is true (compare the definition of 'consequence').

is no interpretation under which all the sentences of Γ are true and ϕ is false.

The set of sentences Γ is *consistent* (or *satisfiable*) if and only if there is an interpretation under which all the sentences of Γ are true.

(Instead of saying that a sentence ϕ is a consequence of $\{\psi\}$, i.e., of the set whose only member is the sentence ψ, we shall usually say simply that ϕ is a consequence of ψ. Similarly, instead of saying that $\{\phi\}$ is consistent, or that $\{\phi, \psi\}$ is consistent, we shall usually say that ϕ is consistent, or that ϕ is consistent with ψ.)

Now these definitions, while they tell us what the crucial terms 'valid', 'consequence', and 'consistent' *mean* (relative to the artificial language $\mathfrak{L}$, of course), obviously do not enable us to decide immediately which sentences are valid and which not, which are consequences of which, or which sets of sentences are consistent. They only give us a basis with reference to which we may proceed to investigate the problem of how to make decisions of the kinds just mentioned.

It will be seen that our definition of 'valid' amounts roughly to this: a sentence ϕ is valid if and only if ϕ comes out true no matter what non-empty set is chosen as the universe of discourse and no matter how we assign appropriate entities to the non-logical constants of $\mathfrak{L}$. (For individual constants, the 'appropriate entities' are elements of the set, for *n*-ary predicates they are *n*-ary relations among elements of the set, and for sentential letters they are truth-values.) Thus, to show that a given sentence is *not* valid it suffices to give a non-empty set $\mathfrak{D}$ and an assignment of appropriate entities to the non-logical constants of $\mathfrak{L}$ which makes ϕ false. In effect, this is the same as giving in relevant detail an interpretation $\mathfrak{I}$ such that ϕ is false under $\mathfrak{I}$.

Here are some examples of valid sentences:

$$(x)Fx \rightarrow Fa$$
$$Fa \rightarrow ((x)Fx \rightarrow Fa)$$
$$(x)(Fx \vee (Gx \,\&\, Hx)) \leftrightarrow ((x)(Fx \vee Gx) \,\&\, (x)(Fx \vee Hx))$$
$$(P \rightarrow (\exists y)(x)(Fxy \vee -Fxy))$$

For further examples, see the list of theorems of logic in Chapter 7, section 5.

Examples of sentences that are not valid:

$$Fa$$
$$(\exists x)Fx$$
$$(Fa \rightarrow Ga) \rightarrow (-Fa \rightarrow -Ga)$$
$$(x)Fx \vee (x)-Fx$$
$$(x)(\exists y)Fxy \rightarrow (\exists y)(x)Fxy$$

For each of these sentences it is easy to give an interpretation under which the given sentence is false.

Examples of consequence (in each case, the last sentence of the group is a consequence of the set consisting of the remaining sentences of the group):

(1) $(x)(Fx \rightarrow Gx)$
$(x)(Gx \rightarrow Hx)$
$(x)(Fx \rightarrow Hx)$

(2) $(x)(Fx \rightarrow Gx)$
Fa
Ga

(3) $(x)Fx$
$(\exists x)Fx$

(4) $(x)(Fx \rightarrow (\exists y)(Gy \,\&\, Hxy))$
$(y)(Gy \rightarrow -Hay)$
$-Fa$

(5) $Fa \,\&\, -Fa$
Hb

Examples of consistency (each group is a consistent set of sentences):

(1) $(x)(Fx \rightarrow Gx)$
$-(\exists x)Gx$

(2) $(x)(Fx \rightarrow (\exists y)(Gy \,\&\, Hxy))$
$(x)(y)(Gy \rightarrow -Hxy)$

(3) $(x)(Fx \vee Gx)$

(4) $(\exists x)Fx$

The following generalizations, which may be established on the basis of our various definitions, will help to illuminate the ideas we have been discussing and will provide some evidence that the formal concepts as defined have properties analogous to those of the intuitive concepts to which they correspond.

1) For any sentence ϕ, if ϕ is a consequence of the set of sentences Γ and each sentence in Γ is a consequence of the set of sentences Δ, then ϕ is a consequence of Δ.

2) For any sentence ϕ, ϕ is a consequence of the empty set of sentences if and only if ϕ is valid.

3) For any sentences $\phi, \psi_1, \psi_2, \ldots, \psi_n$, ϕ is a consequence of $\{\psi_1, \psi_2, \ldots, \psi_n\}$ if and only if the sentence $((\ldots(\psi_1 \,\&\, \psi_2) \,\&\, \ldots \,\&\, \psi_n) \rightarrow \phi)$ is valid.

4) For any sentence ϕ, if ϕ is a member of the set of sentences Γ, then ϕ is a consequence of Γ.

5) For any sentence ϕ, if ϕ is a consequence of a set of sentences Γ, then ϕ is a consequence of any set of sentences that includes Γ as a subset.

6) For any sentence ϕ, ϕ is a consequence of the sentences Γ together with a sentence ψ if and only if $(\psi \rightarrow \phi)$ is a consequence of Γ alone.

7) If two interpretations $\mathfrak{I}$ and $\mathfrak{I}'$ have the same domain and agree in what they assign to all non-logical constants occurring in a given sentence ϕ, then ϕ is true under $\mathfrak{I}$ if and only if ϕ is true under $\mathfrak{I}'$.

8) If a sentence ϕ is like ϕ' except for having the individual constants $\gamma_1, \gamma_2, \ldots, \gamma_n$ wherever ϕ' has, respectively, the distinct (i.e., no two are identical) individual constants $\beta_1, \beta_2, \ldots, \beta_n$, and if an interpretation $\mathfrak{I}'$ is like an interpretation $\mathfrak{I}$ except that it assigns to $\beta_1, \beta_2, \ldots, \beta_n$, respectively, what $\mathfrak{I}$ assigns to $\gamma_1, \gamma_2, \ldots, \gamma_n$, then ϕ is true under $\mathfrak{I}$ if and only if ϕ' is true under $\mathfrak{I}'$.

9) For any formula ϕ, variable α, and individual constant β: if $\phi\ \alpha/\beta$ is a sentence, then it is a consequence of $(\alpha)\phi$.

10) For any formula ϕ, variable α, and individual constant β: if $\phi\ \alpha/\beta$ is a consequence of the set of sentences Γ and if β occurs neither in ϕ nor in any member of Γ, then $(\alpha)\phi$ is a consequence of Γ.

11) For any formula ϕ and variable α: if $(\exists\alpha)\phi$ is a sentence, then it is a consequence of $-(\alpha)-\phi$, and vice versa.

12) For any formula ϕ, variable α, and individual constant β: if $(\exists\alpha)\phi$ is a sentence, then it is a consequence of $\phi\ \alpha/\beta$.

13) For any sentence ϕ, formula ψ, variable α, and individual constant β: if $(\psi\ \alpha/\beta \rightarrow \phi)$ is a consequence of the set of sentences Γ and if β occurs neither in ϕ, ψ, nor in any member of Γ, then $((\exists\alpha)\psi \rightarrow \phi)$ is a consequence of Γ.

14) For any sentence ϕ, if ϕ is a consequence of the set of sentences Γ and each sentence in Γ is valid, then ϕ is valid.

15) For any formula ϕ, variable α, and individual constant β not occurring in ϕ: $\phi\ \alpha/\beta$ is valid if and only if $(\alpha)\phi$ is valid.

16) For any sentence ϕ, ϕ is valid if and only if $-\phi$ is not satisfiable.

17) For any sentence ϕ and set of sentences Γ (possibly infinite), ϕ is a consequence of Γ if and only if there is a finite subset Γ' of Γ such that ϕ is a consequence of Γ'.

18) If a sentence $(\alpha)\psi$ is like a sentence $(\alpha')\psi'$ except that it has the variable α where and only where $(\alpha')\psi'$ has the variable α', then $(\alpha)\psi$ is a consequence of $(\alpha')\psi'$ and vice versa.

19) Let $\mathfrak{I}$ be an interpretation, $\mathfrak{D}$ its domain, and Δ a set of individual constants. If each element of $\mathfrak{D}$ is assigned by $\mathfrak{I}$ to at least one constant in Δ, and if $\phi\ \alpha/\beta$ is true under $\mathfrak{I}$ for all constants β in Δ, then $(\alpha)\phi$ is true under $\mathfrak{I}$.

At this stage, of course, it is not expected that the student will be able to establish all of the foregoing assertions. But he should be able to understand them, and they will give him a better intuitive grasp of the concepts defined in this chapter. Proof of some of them—e.g., 1–6, 11, 14, 16 and 18—is easy. For instance, 2 is established by the following argument:

Let ϕ be any sentence, and suppose that ϕ is a consequence of Λ (the empty set of sentences). Now, let $\mathfrak{I}$ be any interpretation whatever. Clearly and trivially, all members of Λ are true under $\mathfrak{I}$, since Λ has no members. Hence ϕ is true under $\mathfrak{I}$, since ϕ is a consequence of Λ. Thus ϕ is true under every interpretation, i.e., it is valid. On the other hand, suppose that ϕ is valid. Then ϕ is true under every interpretation; hence there is no interpretation under which all members of Λ are true while ϕ is not true. Therefore, ϕ is a consequence of Λ.

For another example, consider 6, the so-called deduction theorem. Let ϕ and ψ be sentences, and Γ a set of sentences. Now ϕ is a consequence of Γ together with ψ if and only if there is no interpretation under which all the sentences of Γ are true and ψ is true and ϕ is false. But this holds if and only if there is no interpretation under which all the sentences of Γ are true and $(\psi \rightarrow \phi)$ is false, and this in turn says that $(\psi \rightarrow \phi)$ is a consequence of Γ. Hence, ϕ is a consequence of Γ together with ψ if and only if $(\psi \rightarrow \phi)$ is a consequence of Γ alone.

Item 7, though it is perhaps intuitively obvious, is best proved by a method that will not be discussed until later. The same is true of the fundamentally important theorem 8. From 8 it is not difficult to obtain 9, 10, 15, and 19. Items 12 and 13 are established by our proofs of derived rules EG and ES, in Chapter 7. Item 17 follows as a corollary from the fact that if ϕ is a consequence of Γ, then ϕ is derivable from Γ by the rules given in Chapter 7.

Now that we have discussed interpretation and truth we are in a better position to investigate the semantic relationships between the artificial language $\mathfrak{L}$ and the natural languages used in everyday life. This is the topic of the next chapter.

EXERCISES

1. Show that if $-\phi$ is a valid sentence, then every sentence is a consequence of ϕ.
2. Similarly, show that if $\{\phi, \psi\}$ is an inconsistent set of sentences, then every sentence is a consequence of it.
3. Show that if ϕ is a valid sentence, then ϕ is a consequence of every set of sentences.
4. Show that if ϕ and ψ are sentences, then ψ is a consequence of ϕ if and only if $(\phi \rightarrow \psi)$ is valid.
5. Show that if $-\phi$ is a valid sentence, then ϕ is inconsistent with itself.
6. Give an interpretation $\mathfrak{I}$ such that '*Fa*' is true under every '*a*'-variant of $\mathfrak{I}$.
7. Referring to exercise 1, Chapter 3,
   (a) give an interpretation under which (3) is true.
   (b) give an interpretation under which (3) is false.
   (c) give an interpretation under which (6) is true.
   (d) give an interpretation under which (10) is true.

8. Referring to the same exercise, explain why there cannot be an interpretation under which (6) is false. Do the same for (10).
9. Prove or disprove: For any sentences ϕ, ψ, χ: if ϕ is a consequence of ψ or of χ, then ϕ is a consequence of $(\psi \vee \chi)$.
10. In each of the following cases, give an interpretation under which the last sentence listed is false and the remaining sentences are true (thus showing that the last sentence is not a consequence of the remaining sentences).
    (a) $Ga \rightarrow Fa$
    $-Ga$
    $-Fa$
    (b) $(x)(Hx \rightarrow Gx)$
    $(x)(Fx \rightarrow Gx)$
    $(\exists x)(Fx \,\&\, Hx)$
    (c) $(\exists x)Fx$
    $(\exists x)Gx$
    $(\exists x)(Fx \,\&\, Gx)$
    (d) $(x)(\exists y)Fxy$
    $(\exists y)(x)Fxy$
    (e) $(x)(Fx \vee Gx)$
    $(x)Fx \vee (x)Gx$
    (f) $Ga \rightarrow Fa$
    Fa
    Ga
    (g) $(x)Fx \rightarrow Ga$
    $(x)(Fx \rightarrow Ga)$
    (h) $(P \rightarrow (Q \rightarrow R))$
    $(P \rightarrow (R \rightarrow Q))$
11. Show that the following sets of sentences are consistent.
    (a) $(x)(\exists y)Fxy$
    $(x)(Gx \rightarrow (\exists y)Fyx)$
    $(\exists x)Gx$
    $(x)-Fxx$
    (b) $(x)(Px \vee Qx) \rightarrow (\exists x)Rx$
    $(x)(Rx \rightarrow Qx)$
    $(\exists x)(Px \,\&\, -Qx)$
    (c) $(x)-Fxx$
    $(x)(y)(z)((Fxy \,\&\, Fyz) \rightarrow Fxz)$
    $(x)(\exists y)Fxy$
12. Give an inconsistent set that contains exactly five sentences and is such that every proper subset is consistent.
13. How many elements must there be in the domain of any interpretation under which the following sentence is true?

    $$Fa \,\&\, -Fb.$$

    Write a sentence that is satisfiable and will be true only under interpretations having at least three elements in their domains.
    Write a sentence that is satisfiable and will be true only under interpretations having infinitely many elements in their domains. (Hint: look at item (c) of exercise 11.)
14. Explain why there cannot be an interpretation under which the following sentence is false:

    $$(P \rightarrow Q) \vee (Q \rightarrow R).$$

5

TRANSLATING THE NATURAL LANGUAGE INTO $\mathfrak{L}$

The next problem to occupy our attention is that of translating sentences of the natural language into those of the artificial language $\mathfrak{L}$. Since the sentences of $\mathfrak{L}$ are not true or false absolutely, but only relative to a given interpretation, it is clear that the task of finding translations for particular sentences of the natural language must likewise be relativized to an interpretation. In the course of considering a number of examples we endeavor to make clear that even specifying an interpretation does not suffice for making the problem of translation determinate; different ways of giving the same interpretation lead to different translations of the same natural language sentence. Consequently, for purposes of translation we have to establish a standard way of giving interpretations; this may be done by means of the notion of 'English predicate'. Nevertheless, the task of translating the natural language into our artificial language is still one for which it is practically impossible to find systematic rules. For, unlike the language $\mathfrak{L}$, the natural language does not have a simple and regular grammar; on the whole we learn it by practice and not by precept. The meaning of its expressions depends heavily on context, whereas relative to a given interpretation each expression of $\mathfrak{L}$ always and everywhere denotes exactly the same thing.

In practice, the following advice proves much more useful than its obvious deficiencies would seem to permit: to translate a sentence of the natural language into the artificial language $\mathfrak{L}$, ask yourself what the natural language sentence means, and then try to find a sentence of $\mathfrak{L}$ which, relative to a standard specification of an interpretation, has as nearly as possible that same meaning.

1. *Introduction.* As we saw in Chapter 4, each interpretation of $\mathfrak{L}$ determines truth-conditions for all the sentences of $\mathfrak{L}$. Thus, by applying the definition of 'true under $\mathfrak{I}$' we can establish such statements as these: For any interpretation $\mathfrak{I}$,

(1) 'Ds' is true under $\mathfrak{I}$ if and only if the object that $\mathfrak{I}$ assigns to 's' belongs to the set that $\mathfrak{I}$ assigns to 'D';
(2) '$-Ds$' is true under $\mathfrak{I}$ if and only if the object that $\mathfrak{I}$ assigns to 's' does not belong to the set that $\mathfrak{I}$ assigns to 'D';
(3) '$(\exists x)Dx$' is true under $\mathfrak{I}$ if and only if some element of the domain of $\mathfrak{I}$ belongs to the set that $\mathfrak{I}$ assigns to 'D';
(4) '$Ds \rightarrow (\exists x)Dx$' is true under $\mathfrak{I}$ if and only if either the object that $\mathfrak{I}$ assigns to 's' does not belong to the set that $\mathfrak{I}$ assigns to 'D', or some element of the domain of $\mathfrak{I}$ belongs to the set that $\mathfrak{I}$ assigns to 'D', or both.

and so on. If a particular interpretation $\mathfrak{I}$ is given in such a way that the following can be shown (without using additional factual information):

The set of human beings is the domain of $\mathfrak{I}$,
Socrates is the object that $\mathfrak{I}$ assigns to 's',
The set of all persons who died in 399 B.C. is the set that $\mathfrak{I}$ assigns to 'D',

we can obtain as further consequences of the definition:

(1′) 'Ds' is true under $\mathfrak{I}$ if and only if Socrates died in 399 B.C.;
(2′) '$-Ds$' is true under $\mathfrak{I}$ if and only if Socrates did not die in 399 B.C.;
(3′) '$(\exists x)Dx$' is true under $\mathfrak{I}$ if and only if someone died in 399 B.C.;
(4′) '$Ds \rightarrow (\exists x)Dx$' is true under $\mathfrak{I}$ if and only if, if Socrates died in 399 B.C., then someone died in 399 B.C. (where 'if ... then' is to be understood in the sense of 'not ... or').

In fact, if the interpretation $\mathfrak{I}$ were fully given in this way, we could in principle obtain for any sentence ϕ of $\mathfrak{L}$ an appropriate instance of the schema

(T) X is true under $\mathfrak{I}$ if and only if p,

where 'X' is replaced by a name of ϕ and 'p' is replaced by some sentence S of the natural language. The sentence S will state in English the conditions under which ϕ is true, relative to the interpretation $\mathfrak{I}$, and to that extent S and ϕ share the same meaning.

Now our task in the present chapter may be characterized roughly as that of finding a way of *translating* English sentences into the formalized language $\mathfrak{L}$. Given an English sentence, we want a way of producing a formal sentence that has as nearly as possible the same meaning.

As is immediately evident, this demand has no sense whatever as long as the language $\mathfrak{L}$ is uninterpreted. The bare minimum that can be expected of translation is that true sentences should be translated by true sentences and false by false, but no sentence of $\mathfrak{L}$ has any truth-value at all except relative to an interpretation. Once an interpretation $\mathfrak{I}$ is given, however, it may be possible, for a particular English sentence S, to find a formal sentence ϕ that qualifies at least partially as a translation of S into the language $\mathfrak{L}$.

There is admittedly a great deal of vagueness in the foregoing, but before further analysis it will be instructive to look at some examples of the sort of translation expected in practice. Let the interpretation $\mathfrak{I}$ be given as on pages 58-9 above, and consider the sentences '2 is even', '$2 < 3$', '$2 + 3 = 5$', and '2 is not odd'. Their translations will be, respectively, 'E^1a_2', '$L^2a_2a_3$', '$S^3a_2a_3a_5$', and '$-O^1a_2$'. The corresponding instances of schema (T), namely

'E^1a_2' is true under $\mathfrak{I}$ if and only if 2 is even,
'$L^2a_2a_3$' is true under $\mathfrak{I}$ if and only if $2 < 3$,
'$S^3a_2a_3a_5$' is true under $\mathfrak{I}$ if and only if $2 + 3 = 5$,
'$-O^1a_2$' is true under $\mathfrak{I}$ if and only if 2 is not odd,

may be established on the basis of the interpretation as given, together with clauses 2 and 3 of the definition of 'true under $\mathfrak{I}$'.

As might be expected, more complicated cases are more difficult to deal with. Consider some further examples:

(1) Every prime is odd.
$(x)(Px \rightarrow Ox)$

Here it would have to be understood as a background assumption that we are speaking of positive integers. Clearly the given formal sentence would equally well translate 'all primes are odd' and 'a prime is always odd'. If we do not rely on the background assumption, we get 'every prime positive integer is odd', 'all prime positive integers are odd', 'every positive integer that is prime is odd', 'all positive integers that are prime are odd', 'a positive integer is odd if it is a prime', 'any prime positive integer is odd', 'whatever positive integer you take, either it is not prime or it is odd, or both', and many other possible renditions of the same idea. In general, something of the form

All A is B

will be representable by a corresponding sentence of the form

$$(\alpha)(\phi \rightarrow \psi).$$

It must be remembered, of course, that the latter will be true under $\mathfrak{I}$ if $(\alpha) - \phi$ or $(\alpha)\ \psi$ is true under $\mathfrak{I}$, although it is doubtful that the former will be regarded as true whenever the set corresponding to *A* is empty or that corresponding to *B* is the whole universe of discourse.

Compare also

(2) Every positive integer is odd.
$(x)Ox$

Thus, if the set *A* happens to be the whole universe of discourse, the sentence

All *A* is *B*

gets a somewhat simpler translation.

(3) Some primes are odd.
$(\exists x)(Px \,\&\, Ox)$

In general, a sentence of the form

Some *A* is *B*

will be representable by a corresponding sentence of the form

$$(\exists\alpha)(\phi \,\&\, \psi).$$

Note that it would not do to use

$$(\exists\alpha)(\phi \rightarrow \psi),$$

which is much too weak, being equivalent to

$$(\exists\alpha)(-\phi \vee \psi),$$

and hence to

$$(\exists\alpha)-\phi \vee (\exists\alpha)\psi.$$

(4) Some positive integers are odd.
$(\exists x)Ox$

(5) No primes are odd.
$(x)(Px \rightarrow -Ox)$ or $-(\exists x)(Px \,\&\, Ox)$

In the following, 'integer' is used in the sense of 'positive integer'.

(6) No integer is both even and odd.
$-(\exists x)(Ex \,\&\, Ox)$

(7) 1 is less than every integer.
$(x)La_1x$

(8) 1 is less than every other integer.
$(x)(-Ia_1x \rightarrow La_1x)$

(9) For every integer there is a greater integer.
$(x)(\exists y)Lxy$

In expressing 'greater than' by means of a predicate that would properly translate 'less than' we give an example of how translation depends upon the available vocabulary. Of course, if we had a binary predicate '*G*' that represented 'greater than' relative to $\mathfrak{I}$, we could translate our sentence as

$$(x)(\exists y)Gyx,$$

but with the predicates given we can do no better than

$$(x)(\exists y)Lxy.$$

The difference may seem small, but its existence makes itself felt if, using only the predicate '*L*', we try to translate "For any two positive integers, if the first is greater than the second, then the second is less than the first".

(10) There is an integer that is greater than every integer.
$(\exists y)(x)Lxy$

(11) There is an integer that is greater than every integer other than itself.
$(\exists y)(x)(-Ixy \rightarrow Lxy)$

(12) If the product of two integers is even, then at least one of them is even.
$(x)(y)(z)((Mxyz \,\&\, Ez) \rightarrow (Ex \vee Ey))$

(13) The product of two even integers is always a multiple of 4.
$(x)(y)(z)(((Ex \,\&\, Ey) \,\&\, Mxyz) \rightarrow (\exists w)Mwa_4z)$

(14) Every even integer greater than 4 is the sum of two primes.
$(x)((Ex \,\&\, La_4x) \rightarrow (\exists y)(\exists z)(Syzx \,\&\, (Py \,\&\, Pz)))$

(15) Every integer greater than 1 is (evenly) divisible by some prime.
$(x)(La_1x \rightarrow (\exists y)(\exists z)(Py \,\&\, Myzx))$

(16) There is no greatest prime.
$(x)(Px \rightarrow (\exists y)(Py \,\&\, Lxy))$

(17) For every pair of primes differing by 2, there is a pair of greater primes differing by 2.
$(x)(y)(((Px \,\&\, Py) \,\&\, (Sxa_2y \vee Sya_2x)) \rightarrow$
$(\exists z)(\exists w)(((Lxz \,\&\, Lyw) \,\&\, (Pz \,\&\, Pw)) \,\&\, (Sza_2w \vee Swa_2z)))$

(18) For every prime of the form $p^2 + 1$ there is a greater prime of that form.

$$(x)((Px \,\&\, (\exists y)(\exists z)(Myyz \,\&\, Sa_1zx)) \rightarrow$$
$$(\exists w)((Pw \,\&\, Lxw) \,\&\, (\exists y)(\exists z)(Myyz \,\&\, Sa_1zw)))$$

Though it would be a tedious job, we could demonstrate in each of these cases, too, the corresponding instance of schema (T) above.

For a different group of examples, let $\mathfrak{I}'$ be an interpretation having the set of all characters in *David Copperfield* as its domain (we take for granted that the set is not empty) and assigning denotations to non-logical constants as follows:

I^1: the set of those who intervene
W^1: the set of those who are willing
P^1: the set of those who will throw off their pecuniary shackles
M^2: the binary relation that holds between characters $\mathfrak{m}$ and $\mathfrak{n}$ when $\mathfrak{m}$ will marry $\mathfrak{n}$
F^2: the binary relation that holds between characters $\mathfrak{m}$ and $\mathfrak{n}$ when $\mathfrak{m}$ is a friend of $\mathfrak{n}$
L^2: the binary relation that holds between characters $\mathfrak{m}$ and $\mathfrak{n}$ when $\mathfrak{m}$ will lend money to $\mathfrak{n}$
a: Agnes m: Micawber
b: Barkis p: Peggotty
h: Heep

On this basis we have the following:

(19) Heep is willing but Agnes is not.
$Wh \,\&\, -Wa$

(20) If Barkis is willing, Peggotty will marry him.
$Wb \rightarrow Mpb$

This example illustrates a problem that commonly arises, especially in connection with non-mathematical sentences. The two clauses of our English sentence carry an implicit reference to time, and not just to any time, but to *related* times. The sense is at least that if Barkis is willing at some time t (perhaps the present), then Peggotty will marry him at some time t' later than t. That our formula does not do justice to this may be seen from the fact that, although

$$-Mpb \rightarrow -Wb$$

is a consequence of it, the English sentence

> If it is not the case that Peggotty will marry Barkis, then he is not willing

would not be thought to follow from (20).

The moral here, the student may feel, is that we should not attempt to translate ordinary sentences of this sort unless the interpretation allows us to make the appropriate references to time. But a little reflection will show that even when such references are possible the essential difficulty remains. After all, the idea is not merely that for any time, either Barkis is not willing at that time or Peggotty will marry him at a somewhat later time, but rather that there is a *causal* connection involved. Whether, using only quantifiers and truth-functional connectives, we can adequately and compactly express this sort of connection, is a question that quickly leads to rather deep philosophical issues. For the present we shall plead that, as a practical task, translation is a matter of doing the best one can with the means available.

(21) If Barkis is willing, then Peggotty will marry him unless someone intervenes.
$Wb \rightarrow (-(\exists x)Ix \rightarrow Mpb)$

(22) If no one intervenes, then either Agnes will marry Heep or Heep will not lend money to Barkis.
$-(\exists x)Ix \rightarrow (Mah \vee -Lhb)$

(23) Micawber will throw off his pecuniary shackles only if some friend of his will lend him money.
$Pm \rightarrow (\exists x)(Fxm \,\&\, Lxm)$

(24) Heep will not lend money to anyone who is a friend of Micawber.
$(x)(Fxm \rightarrow -Lhx)$

(25) Although everybody is willing, nobody will marry anybody unless Micawber throws off his pecuniary shackles.
$(x)Wx \,\&\, (-Pm \rightarrow (x)(y)-Mxy)$

(26) Barkis will lend money to Micawber if Heep does not intervene, and whoever intervenes is no friend of Micawber's.
$(-Ih \rightarrow Lbm) \,\&\, (x)(Ix \rightarrow -Fxm)$

(27) Everyone has some friend who will borrow money from a friend of Micawber's unless Micawber intervenes.
$(x)(\exists y)(Fyx \,\&\, (\exists z)(Fzm \,\&\, (-Im \rightarrow Lzy)))$

Again, for each of these pairs the corresponding instance of schema (T) could be demonstrated.

2. *Interpretation and translation.* The extent to which an interpretation of $\mathfrak{L}$ gives meaning to the sentences of $\mathfrak{L}$ will be clear if we bear in mind Frege's distinction between the sense and the denotation of linguistic

expressions. Given the sense, the denotation (if any) is thereby fixed, but the converse does not hold. Therefore, although an interpretation always assigns a denotation to each predicate, individual constant, and sentence of $\mathfrak{L}$ it does not thereby determine a sense. Of course, by assigning a denotation to an expression, it puts a limit on the range of possible senses that the expression could have. Thus if, under an interpretation $\mathfrak{I}$, the individual constant 'a_2' denotes 2 it can have the sense of 'the only even prime' or that of 'the positive square root of 4', but not that of 'the positive square root of 9'. This, however, does not suffice for our purposes. In order to decide whether a formal sentence ϕ is or is not a satisfactory translation of an English sentence *S*, we must in some way associate a sense with ϕ, and not merely a denotation.

It is probably evident that we have in fact been doing this tacitly. We have required that in order for the formal sentence ϕ to qualify as a translation of *S* relative to the interpretation $\mathfrak{I}$ it must at least be possible to establish the corresponding instance of schema (T) on the basis of our definition of 'true under $\mathfrak{I}$', together with various synonymies in the natural language. It is essential to note that in establishing instances of (T) we utilize the *way* in which $\mathfrak{I}$ is described; in other words, if the same interpretation $\mathfrak{I}$ is 'given' in different ways we may obtain very different English sentences corresponding to the same sentence of the artificial language. For example, suppose that the interpretation $\mathfrak{I}$ that was described on pages 58-9 above had been described or 'given' in the same way except that instead of

$$a_2\text{: } 2$$

we had

$$a_2\text{: the only even prime.}$$

Then, instead of

'$L^2a_2a_3$' is true under $\mathfrak{I}$ if and only if 2 is less than 3,

which we obtained from the definition of 'true under $\mathfrak{I}$' together with the former description of $\mathfrak{I}$, we would have

'$L^2a_2a_3$' is true under $\mathfrak{I}$ if and only if the only even prime is less than 3.

Thus we should be led to regard '$L^2a_2a_3$' as a translation of the English sentence 'the only even prime is less than 3', whereas formerly, relative to the same interpretation $\mathfrak{I}$, we regarded it as a translation of the English sentence '2 is less than 3'. We see, therefore, that by our procedure the question whether a formal sentence ϕ is an adequate translation of an English sentence *S* is relativized not merely to an interpretation $\mathfrak{I}$, but even to a particular description or way of 'giving' the interpretation $\mathfrak{I}$.

It is obvious that in *giving* an interpretation one must somehow indicate which objects are assigned to which constants, and a natural way of doing this is to supply, for each constant under consideration, an English expression denoting the object to be assigned to that constant. Thus, in giving the numerical interpretation $\mathfrak{I}$ discussed above, we use a name or description of the number 2 to indicate which object is assigned to 'a_2', and we indicate the relation assigned to 'L^2' by using some such phrase as 'is less than'. But these phrases not only have denotations; they also have senses. We have been making use of this fact by tacitly assuming that the English expressions employed in giving an interpretation indicate not only the denotations of the non-logical constants involved, but also their senses. Thus, we have been assuming that not only is

$$L^2a_2a_3$$

true if and only if the only even prime is less than 3, but it *means* that the only even prime is less than 3.

Problems of translation will therefore take on a more determinate character if, when we ask for translations, we specify the relevant interpretation in a standard way. Let that standard way be as follows: the domain is to be indicated by giving a name or a description of a non-empty set; for the non-logical constants under consideration denotations are to be indicated by giving (a) English sentences for sentential letters, (b) English names or descriptive phrases for individual constants, and (c) English predicates for predicate constants. We borrow the notion of 'English predicate' from Quine: an English predicate is like an English sentence except that it contains the counter '①', or the counters '①' and '②', or '①', '②', and '③', etc. in one or more places where names or descriptive phrases occur directly. Thus, each of these is an English predicate:

① < ②

② < ① or ① < ③

If ② is a brother of ①'s best friend, then Frank and ② are somehow related;

but this is not a predicate:

① wonders whether ② is coming to dinner,

nor is this:

① < ③.

Of the counters occurring in an English predicate, the one with the highest numeral determines the *degree* of the predicate, and in giving an interpretation it is of course necessary that the English predicate have the same degree as that of the formal predicate which it interprets.

Illustrating these conventions we give the interpretation $\mathfrak{I}$ of pages 58–9 in our standard way:

$\mathfrak{D}$: the set of positive integers
E^1: ① is even
O^1: ① is odd
P^1: ① is prime
L^2: ① < ②
I^2: ① = ②
S^3: ① + ② = ③
M^3: ① · ② = ③
a_1: 1 P: 1 = 1
a_2: 2 Q: 1 ≠ 1
a_3: 3 R: 1 = 1
. .
. .

For purposes of translating English sentences relative to $\mathfrak{I}$, the entries in this table now have a double import. The entry

S^3: ① + ② = ③,

for example, indicates (a) that the interpretation $\mathfrak{I}$ associates with 'S^3' the set of all triples of positive integers satisfying the English predicate (which, of course, is in effect a sentence-form) '① + ② = ③', and (b) that the sense of 'S^3' is that of this English predicate.

3. *Translating connectives and quantifiers*. Up to this point we have been discussing primarily the interpretation of the non-logical constants of $\mathfrak{L}$. Let us now consider the connectives and quantifiers. These, as will be seen from the definition of 'true under $\mathfrak{I}$', do not behave in quite the same way as their counterparts in the natural language. This is due partly, but only partly, to the vagueness and ambiguity of those counterparts. Perhaps the most important differences arise from the fact that we interpret the connectives of $\mathfrak{L}$ truth-functionally. That is to say, if a sentence of $\mathfrak{L}$ is built up out of other sentences by means of the connectives, then its truth-value under any interpretation depends solely on (i.e., is a function of) the truth-values which that interpretation gives to the components. Since such words as 'or', 'and', 'if, ... then' are not always (if ever) used truth-functionally in everyday language, their representation by the connectives of $\mathfrak{L}$ is always more or less questionable.

The differences are perhaps at a minimum in the case of '&' and 'and'. Given an interpretation $\mathfrak{I}$ and sentences ϕ, ψ, the sentence $(\phi \,\&\, \psi)$ is true under $\mathfrak{I}$ just in case both conjuncts are true under $\mathfrak{I}$; otherwise it is false. This corresponds reasonably well with the use of 'and' in the natural language, although sometimes the latter has the sense of 'and then', so that 'he brushed his teeth and went to bed' is quite different from 'he went to bed and brushed his teeth'. Observe also that 'and' is sometimes used between names, as in the sentence '2 and 4 are even numbers'; in such cases we usually render the sentence by a conjunction of two or more sentences of $\mathfrak{L}$: in this case, by

$$Ea_2 \,\&\, Ea_4.$$

(But in some cases, e.g., 'David and Agnes were married', this sort of transformation would change the sense radically.)

The connective 'v' raises another problem. A disjunction is true under a given interpretation $\mathfrak{I}$ just in case one or both of the disjuncts is true under $\mathfrak{I}$. Thus, as far as the sense of 'v' is concerned, it is perfectly possible to have a true disjunction in which both disjuncts are true. Under our numerical interpretation $\mathfrak{I}$, the following would be such a case:

$$La_1a_2 \vee La_1a_2.$$

Of course, there will also be true disjunctions in which it is *not* possible for both disjuncts to be true; e.g.,

$$La_1a_2 \vee -La_1a_2.$$

But in cases like this the possibility of joint truth is excluded, not by the sense of 'v', but by those of 'La_1a_2' and '$-La_1a_2$'. Since the English word 'or' nearly always connects sentences the joint truth of which is excluded on the basis of their content or on the basis of certain background assumptions, it is hard to decide whether this word occasionally has an 'exclusive' sense, as some logicians have asserted, or always has an 'inclusive' sense analogous to that of 'v'.

If one considers the connective '$\rightarrow$' as somehow corresponding to the words 'if ... then', there are several important differences that must be borne in mind. We use this connective in the so-called material sense: a conditional is true under a given interpretation $\mathfrak{I}$ if either the antecedent is false under $\mathfrak{I}$ or the consequent is true under $\mathfrak{I}$, or both. Thus, relative to our numerical interpretation,

$$Oa_2 \rightarrow Ea_2,$$
$$Oa_2 \rightarrow -Ea_2,$$
$$-Oa_2 \rightarrow Ea_2$$

are all true. So are

$$(x)((Ex \,\&\, Ox) \rightarrow Ex),$$
$$(x)((Ex \,\&\, Ox) \rightarrow -Ex),$$
$$(x)(Ex \rightarrow (Ex \vee Ox)),$$
$$(x)(-Ex \rightarrow (Ex \vee Ox)).$$

On the other hand,

$$Ea_2 \to Oa_2$$

is false, as one would hope. Now it is safe to say that in everyday language 'if . . . then' is practically never used in the material sense. But just what its ordinary sense *is* (or ordinary senses *are*) is not easy to make out. It has sometimes been argued that the material sense is a sort of 'least common denominator'. For if an 'if . . . then' sentence is true in any ordinary sense of 'if', its translation with '$\to$' will be true with respect to the appropriate interpretation. But this does not mean that whenever we translate 'if . . . then' by '$\to$' we weaken our assertion, i.e., assert less. For instance, the sentence '$-(P \to Q)$' is considerably too strong as a translation of 'it is not the case that if we stand firm the Russians will back down', as may be seen from the fact that 'P' is a consequence of the one, while no one will suggest that 'we will stand firm' follows from the other. At any rate, it is clear that the truth or falsity of 'if . . . then' sentences usually depends in part upon interrelationships between the senses (and not merely the truth-values) of their components; since the opposite is the case with the connective '$\to$' we can expect only a very loose fit.

As concerns 'not', the principal caveat is that its position in ordinary sentences frequently obscures its logical role. In 'all that glisters is not gold', for example, the meaning is clearly 'not all that glisters is gold', so that (with respect to an obvious interpretation) its representation would be something like

$$-(x)(G_1x \to G_2x)$$

and not

$$(x)(G_1x \to -G_2x).$$

Again, if we represent 'Jones will certainly win' by 'Wj' we must be careful not to suppose that 'Jones will certainly not win' is to be translated as '$-Wj$', for this will only mean that it is not certain that Jones will win (and perhaps not certain that he won't). Likewise the two sentences 'I don't want him to come' and 'I do want him to come' may neither of them be true (if I don't care whether he comes or not), and hence for them no sentence of $\mathfrak{L}$ and its negation can be adequate translations.

It is worth noting that often a connective or quantifier will appear in a symbolization when its normal counterpart does not appear in the sentence symbolized. Thus '$\to$' and '&' occur in the usual renditions of 'all men are mortal' and 'some men are mortal'; '$-$' appears when one symbolizes 'we have no bananas', 'nothing is colder than ice', 'I will go unless he comes', etc.; a universal quantifier is employed in symbolizing such sentences as 'a Scout is reverent' (but not 'a lady is present', as Quine has mentioned); and so on. In each such case, however, there will be an English sentence that does contain a corresponding connective or quantifier and that coincides in sense with the formal sentence at least as well as the sentence symbolized does.

The natural language contains a number of connectives that are clearly not truth-functional, as well as certain others that must be classified as borderline. The words 'since' and 'because' are examples of the former type. In the sentence

Jones won because Smith was disqualified

the truth-value of the whole is not completely determined by the truth-values of the parts. Though the whole sentence is indeed false if one or both of the component sentences is false, its truth-value is not similarly determined for the case in which both are true. The connectives 'but', 'although', and 'unless', on the other hand, seem to be borderline. Sentences of the form

--- although ...

or

--- but ...

may for our purposes be considered as equivalent to the corresponding sentences of the form

--- and ...,

and similarly we shall consider sentences of the form

--- unless ...

as equivalent to the corresponding sentences of the form

If not ..., then ---.

Finally, there are several points to be made concerning the universal and existential quantifiers. We have indicated earlier that '$(\exists x)$' represents 'there is (in the domain) *at least* one thing such that', and thus it does not correspond quite exactly to 'some'. Remember, too, that successively occurring quantifiers need not refer (speaking loosely here) to distinct objects of the domain. For example, in order for the sentence

$$(x)(y)(Lxy \vee Lyx)$$

to be true under the interpretation used in our earlier examples, it would have to be the case that each positive integer was less than itself. For this sentence has as a consequence

$$Laa \vee Laa,$$

which in turn has as a consequence

$$Laa.$$

It is best read: 'for any integer x and for any integer y ...' and not 'for any *two* integers x and y ...'. Similar observations hold for existential quantifiers. The sentence

$$(\exists x)(\exists y)Ixy$$

does not express the contradiction that there are *two* things which are nevertheless identical with one another, but only the triviality that something is identical with something.

Note also that if in any general sentence $(\alpha)\phi$ or $(\exists\alpha)\phi$ of $\mathfrak{L}$ we replace all occurrences of α by occurrences of a variable not already occurring therein, the result will have the same translations as the original. In item (1) on page 71 above we could have used

$$(y)(Py \rightarrow Oy)$$

or

$$(z)(Pz \rightarrow Oz);$$

all such formulas will have exactly the same truth-conditions and translations.

Certain connectives of the natural language require quantifiers for their adequate translation. Consider the word 'whenever', occurring in sentences like

> Whenever Jones plays his trumpet the neighbors are fit to be tied.

This sentence is of course not a truth-functional compound of

Jones plays his trumpet

and

The neighbors are fit to be tied,

for the latter seems to refer only to the present, while the compound is more general. To catch the force of 'whenever' we must introduce a variable taking moments of time as values. With respect to an interpretation given as follows:

$\mathfrak{D}$: the set of moments of time
J: At time ① Jones plays his trumpet
N: At time ① the neighbors are fit to be tied
. .
. .

we would obtain as a translation the sentence

$$(x)(Jx \rightarrow Nx),$$

and this sentence is general, not molecular.

Sometimes, as was mentioned earlier, the word 'if' itself has the sense of 'whenever', as in

> If the temperature rises above 90 degrees, the classes are dismissed,

and, consequently, a quantifier will be required in these cases, too. Conversely, 'whenever' sometimes has the sense of 'if', as in

> Whenever the square of a number is less than one, the number itself is less than one,

which makes no essential reference to time. And 'always' sometimes indicates mere generality:

> The sum of the angles of a triangle is always equal to two right angles.

Summing up, we note that there are a number of things to keep in mind when 'translating' back and forth between ordinary language and the formal notation. Some of the most important of these are the following.

1) Only relative to an interpretation do the expressions of $\mathfrak{L}$ have denotation, let alone sense; and, consequently, it is futile to attempt to translate or symbolize unless an interpretation has somehow been given.

2) Once an interpretation has been given, a sort of 'standard' translation may be obtained for any formal sentence, via the definition of 'true under $\mathfrak{I}$; but when the same interpretation is given in different ways the indicated procedure can lead to different, and plainly non-synonymous, translations for the same formal sentence. These differences will result not only from the way in which we describe the objects assigned to the non-logical constants of $\mathfrak{L}$, but also from the way in which we may happen to describe the domain. The question of whether a given translation or symbolization is 'correct', then, is best relativized to an interpretation *as given in a certain way*.

3) We cannot expect too simple a correspondence between the form of an ordinary sentence and that of its counterpart in $\mathfrak{L}$. Even sentences of the form

$$S \text{ and } T$$

are not always best represented by sentences of the form

$$(\phi \,\&\, \psi),$$

and sentences of the form

$$\text{If } S \text{ then } T$$

will perhaps only infrequently have counterparts of the form

$$(\phi \to \psi).$$

4) It is possible to have two formal sentences ϕ and ψ that are consequences of one another (let us call such sentences *equivalent*) and yet are not equally good representations of the same sentence of the natural language. Thus, for example, '3 is a prime' is represented properly by

$$Pa_3$$

but not by

$$--Pa_3,$$

even though these two sentences are equivalent. On the other hand, however, it is presumably impossible for two sentences to be adequate symbolizations of the same sentence and yet not be equivalent, unless, of course, the sentence being symbolized is ambiguous.

All of these difficulties make it obvious that to formulate precise and workable rules for symbolizing sentences of the natural language is a hopeless task. In the more complicated cases, at least, we are reduced to giving the empty-sounding advice: ask yourself what the natural language sentence means, and then try to find a sentence of 𝔏 which, relative to the given interpretation, has as nearly as possible that same meaning. With a little practice the student will find that in fact he can do this fairly successfully in a large class of cases. We trust, however, that while he is developing skill at symbolizing he will also develop an acute awareness of the ways in which his symbolic formulas do not do justice to their originals.

EXERCISES

1. Symbolize the following sentences, using the interpretation given.

$\mathfrak{D}$ = the set of human beings
F: ① is the father of ②
M: ① is the mother of ②
H: ① is the husband of ②
S: ① is a sister of ②
B: ① is a brother of ②
m: Mary
h: Harry
b: William
a: Arthur

(a) Everybody has a father.
(b) Everybody has a father and a mother.
(c) Whoever has a father has a mother.
(d) Harry is a father.

(e) Harry is a grandfather.
(f) All grandfathers are fathers.
(g) Harry is a parent.
(h) All fathers are parents.
(i) All grandparents are parents.
(j) Harry is a husband.
(k) Mary is a wife.
(l) Harry and Mary are husband and wife.
(m) All husbands are spouses.
(n) William is Mary's brother-in-law.
(o) Arthur is William's paternal grandfather.
(p) Mary is Arthur's aunt.
(q) Any unmarried aunt is somebody's sister.
(r) No uncle is an aunt.
(s) All brothers are siblings.
(t) Nobody's grandfather is anybody's mother.

2. Symbolize the following sentences, using the interpretation given.

$\mathfrak{D}$ = the set of mountains and hills
H: ① is higher than ②
M: ① is a mountain
I: ① is identical with ②
E: ① is in England
W: ① is in Wales
S: ① is in Scotland
s: Snowdon
b: Ben Nevis

(a) Snowdon is a mountain in Wales.
(b) Ben Nevis is not in England or Wales, but it is higher than Snowdon.
(c) For every mountain in England there is a higher mountain in Scotland.
(d) Some mountain in Scotland is higher than any mountain in England or Wales.
(e) No mountain is higher than itself.
(f) There are no two mountains, whether in England or in Wales, such that each is higher than the other.
(g) Ben Nevis is at least as high as Snowdon.
(h) Every mountain that is at least as high as Ben Nevis is at least as high as Snowdon.
(i) At least one mountain in Scotland other than Ben Nevis is higher than all mountains in England or Wales.
(j) There are at least two mountains in England.
(k) For every mountain in England there are at least two higher mountains in Scotland.
(l) Every mountain that is higher than all the mountains of England is higher than some of the mountains of Wales.
(m) If Snowdon is at least as high as Ben Nevis, then every mountain that is at least as high as Snowdon is at least as high as Ben Nevis.

(n) Snowdon is in England only if Ben Nevis is in Wales.

(o) Although Snowdon is in England only if Ben Nevis is in Wales, it is not the case that Snowdon is in England if Ben Nevis is in Wales.

(p) There are no mountains in England, Scotland, or Wales; Snowdon and Ben Nevis are only hills.

3. Using the interpretation given, symbolize the premises and conclusion of each of the following arguments. In each case indicate whether or not the formal sentence corresponding to the conclusion is a consequence of the formal sentences corresponding to the premises.

$\mathfrak{D}$ = the set of human beings
C: ① is a centerfielder
P: ① is a pitcher
S: ① scores
F: ① is a friend of ②
O: ① flies out to ②
c: Crabb
j: Jones
r: Robinson
s: Samson

(a) Neither Samson nor any friend of Samson scores. Either Samson scores or Jones scores. Therefore, Jones is not a friend of Samson.

(b) Only pitchers fly out to Robinson. Crabb scores only if Samson flies out to Robinson and Robinson is a centerfielder. Crabb scores. Therefore, Samson is a pitcher.

(c) All friends of Samson are friends of Jones. Any friend of Robinson is a friend of Samson. Therefore, if Crabb is a friend of Robinson then somebody is a friend of Jones.

(d) If Samson is a centerfielder then Crabb is a pitcher. If either Robinson or Jones is a centerfielder then Crabb is not a pitcher. Samson is a centerfielder if anyone is. Therefore, if anyone is a centerfielder, Jones is not.

(e) No centerfielder who does not score has any friends. Robinson and Jones are both centerfielders. Any centerfielder who flies out to Jones does not score. Therefore, if Robinson flies out to Jones then Jones is not a friend of Robinson.

4. For each of the following sentences, give a suitable interpretation and symbolize the sentence relative to that interpretation as given; use no sentential letters.

(a) Singers sing if and only if not asked.
(b) Pepper is sold here and at Rome.
(c) The French are good soldiers.
(d) The Romans conquered the Carthaginians.
(e) All members of the board are over 60 years of age.
(f) All members of the board are related by marriage.
(g) Oil is lighter than water.
(h) Water is composed of hydrogen and oxygen.
(i) All students get good grades if they study.
(j) Some students get good grades if they study.

6

TAUTOLOGOUS SENTENCES

In this chapter we give our attention to a particular group of valid sentences, the so-called tautologies. A tautology is a valid sentence the validity of which depends only upon the semantic properties of the connectives (as contrasted with those of the quantifiers). After an introductory section we give exact definitions of the principal concepts involved. Then we describe the traditional truth-table test for deciding whether or not a given SC sentence (see page 50) is tautologous. Next it is shown how sentences that are not SC sentences may be tested through their associated SC sentences. In the last section we introduce certain inference rules by means of which all and only SC tautologies may be obtained as theorems.

1. *Introduction.* Since ancient times logicians have been especially interested in necessary truths that are necessary by virtue of the meanings of sentential connectives alone. These include, for example, such sentences as

(1) All men are mortal or not all men are mortal,
(2) Snow is white or snow is not white,
(3) Adam begat Seth or Adam did not beget Seth,

which are instances of the so-called *Law of Excluded Middle;* and

(4) Not both: all men are mortal and not all men are mortal,
(5) Not both: snow is white and snow is not white,

(6) Not both: Adam begat Seth and Adam did not beget Seth,

which are instances of the *Law of Contradiction;* and, of course, others infinite in number and variety.

The distinguishing characteristic of all such sentences is that their truth follows from the logical properties of 'or', 'and', 'not', etc.; it is quite independent of the sense of 'all' and 'some', as well as of that of terms like 'men' and 'mortal'. Thus in examples (1) and (4) above we may note that the truth-value would remain the same even if 'all' had the sense of 'no', 'men' that of 'electrons', and 'mortal' that of 'positively charged'. (For contrast consider the sentence

If all men are mortal and all Greeks are men, then all Greeks are mortal,

which, though indeed necessary, depends for its necessity upon the sense of the word 'all'.)

The formal counterparts of these natural language sentences will be called 'tautologous sentences' or 'tautologies'. Thus, in a preliminary and somewhat inexact way we may characterize the tautologies of $\mathfrak{L}$ as those valid sentences the validity of which depends only upon the logical properties of the sentential connectives. In other words, they are those sentences the validity of which can be established on the basis of clauses 1–7 of the definition of 'true under $\mathfrak{I}$' (page 60).

For example, we can easily establish the validity of

(7) $$(x)Fx \vee -(x)Fx$$

by noting that any interpretation $\mathfrak{I}$ will give either the value T or the value F to '$(x)Fx$', and, whichever of these it does, it will in any case give the value T to (7). The same sort of argument can be given for

(8) $$-((x)Fx \;\&\; -(x)Fx)$$

and for

(9) $$((x)Fx \rightarrow Fa) \vee (Fa \rightarrow (x)Fx)$$

but not for

(10) $$(x)Fx \rightarrow Fa$$

or

(11) $$Fa \rightarrow (\exists x)Fx.$$

To establish the validity of (10) and (11) we would have to make use of clauses 8 and 9 of the definition of 'true under $\mathfrak{I}$', for in these cases the logical properties of the quantifiers as well as those of the connectives are essential.

2. *Definition of 'tautology'*. Probably it is immediately evident that no atomic sentence is tautologous. Indeed, no atomic sentence is valid, let alone tautologous. And it is also plain that the truth or falsity of a general sentence relative to an interpretation depends upon the significance given the quantifiers by clauses 8 and 9 of the definition of 'true under $\mathfrak{I}$'; thus no general sentence is tautologous. The remaining, molecular, sentences are built up out of atomic and/or general subsentences by means of the sentential connectives. Thus, a tautology is in effect a sentence that 'comes out true' no matter how truth-values are assigned to its atomic and general parts.

To formulate an exact definition of 'tautologous sentence' we utilize the auxiliary notion of a 'normal assignment' of truth-values to the sentences of $\mathfrak{L}$. Intuitively, a normal assignment of truth-values to the sentences of $\mathfrak{L}$ is one which treats the connectives in such a way that statements 1–5 on page 62 hold. Our definition, therefore, runs as follows:

An assignment $\mathfrak{A}$ of the truth-values T and F to all the sentences of $\mathfrak{L}$ is called *normal* if and only if, for each sentence ϕ of $\mathfrak{L}$,

1) $\mathfrak{A}$ assigns exactly one of the truth-values T or F to ϕ, and

2) if $\phi = -\psi$, then $\mathfrak{A}$ assigns T to ϕ if and only if $\mathfrak{A}$ does not assign T to ψ, and

3) if $\phi = (\psi \vee \chi)$ for sentences ψ, χ, then $\mathfrak{A}$ assigns T to ϕ if and only if $\mathfrak{A}$ assigns T to ψ or assigns T to χ, or both, and

4) if $\phi = (\psi \,\&\, \chi)$ for sentences ψ, χ, then $\mathfrak{A}$ assigns T to ϕ if and only if $\mathfrak{A}$ assigns T to ψ and T to χ, and

5) if $\phi = (\psi \rightarrow \chi)$ for sentences ψ, χ, then $\mathfrak{A}$ assigns T to ϕ if and only if $\mathfrak{A}$ assigns F to ψ or T to χ, or both, and

6) if $\phi = (\psi \leftrightarrow \chi)$ for sentences ψ, χ, then $\mathfrak{A}$ assigns T to ϕ if and only if $\mathfrak{A}$ assigns T to both ψ and χ or assigns F to both.

(Thus, if we make any arbitrary assignment of truth-values to the non-molecular sentences of $\mathfrak{L}$, and then extend it to the molecular sentences as well, in accord with clauses 1–6 above, the result will be a normal assignment of truth-values to the sentences of $\mathfrak{L}$; every normal assignment may be generated in this way).

A sentence ϕ is *tautologous* (or is a *tautology*) if and only if it is assigned the truth-value T by every normal assignment of truth-values T and F to the sentences of $\mathfrak{L}$.

Further, a sentence ϕ is a *tautological consequence* of a set of sentences Γ if and only if ϕ is assigned the truth-value T by every normal assignment that assigns the truth-value T to all sentences of Γ.

A set of sentences Γ is *truth-functionally consistent* if and only if there is at least one normal assignment that assigns the truth-value T to all members of Γ.

A set of sentences Γ is *truth-functionally inconsistent* if and only if it is not truth-functionally consistent.

On the basis of the foregoing, it is easy to see that

(a) A sentence ϕ is a tautological consequence of the empty set of sentences if and only if ϕ is tautologous;

(b) A sentence ϕ is a tautological consequence of the set of sentences $\{\psi_1, \psi_2, \ldots, \psi_n\}$ if and only if the conditional sentence $((\ldots(\psi_1 \& \psi_2) \& \ldots \& \psi_n) \rightarrow \phi)$ is tautologous.

In fact, the following somewhat stronger statement holds but is less easy to establish:

(c) A sentence ϕ is a tautological consequence of a (finite or infinite) set of sentences Γ if and only if either

(i) Γ is empty and ϕ is tautologous, or

(ii) there are sentences $\psi_1, \psi_2, \ldots, \psi_n$ which belong to Γ and are such that $((\ldots(\psi_1 \& \psi_2) \& \ldots \& \psi_n) \rightarrow \phi)$ is tautologous.

What statement (c) adds to (a) and (b) is the fact that if a sentence ϕ is a tautological consequence of an infinite set of sentences Γ, then ϕ is a tautological consequence of some finite subset of Γ.

Every tautologous sentence is valid, since the assignment of truth-values to the sentences of $\mathfrak{L}$ by any interpretation $\mathfrak{I}$ is a normal assignment. For the same reason, whenever a sentence ϕ is a tautological consequence of a set of sentences Γ, it is a consequence of the sentences Γ. On the other hand, there are valid sentences that are not tautologous—'$(x)Fx \rightarrow Fa$' is an example—and there are cases of consequence that are not cases of tautological consequence.

3. *Tautologous SC sentences; truth-tables*. Now although we have before us an exact definition of 'tautologous sentence', we have yet to make plain which sentences are tautologous and which are not. A systematic survey is needed, with numerous examples. In making this survey it helps to take note of certain additional facts that follow from the definition given.

Since the tautologous sentences are supposed to be those valid sentences the validity of which does not depend upon the quantifiers, it is clear that a quantifier-free sentence is tautologous if and only if it is valid:

I. For any sentence ϕ, if ϕ does not contain any quantifiers, then ϕ is tautologous if and only if ϕ is valid.

In particular, since all SC sentences are quantifier-free,

I′. For any SC sentence ϕ, ϕ is tautologous if and only if ϕ is valid.

The tautologous SC sentences are of special interest because of the following fact:

II. For any sentence ϕ, ϕ is tautologous if and only if there is a tautologous SC sentence ψ such that ϕ is a substitution instance of ψ;

where 'substitution instance' is defined thus:

A sentence ϕ is a *substitution instance of an SC sentence* ψ if and only if ϕ is the result of replacing sentential letters of ψ by sentences, the same letter to be replaced at all occurrences by the same sentence.

The intuitive correctness of II may be seen by a consideration of examples. If we look at the argument given earlier to establish the tautologousness of

$$(x)Fx \vee -(x)Fx,$$

we see that an exactly parallel argument will establish that of

$$P \vee -P.$$

We have only to replace all references to '$(x)Fx$' by references to 'P'. Likewise, we establish that

$$((x)Fx \rightarrow Fa) \vee (Fa \rightarrow (x)Fx)$$

is a tautology by arguing that any interpretation $\mathfrak{I}$ will assign either the value T or the value F to 'Fa', and if it assigns F to 'Fa', then it assigns T to the right disjunct and hence T to the whole sentence, while if it assigns T to 'Fa' then it assigns T to the left disjunct and hence again T to the whole sentence; and we could construct a parallel argument for

$$(P \rightarrow Q) \vee (Q \rightarrow P).$$

In general, whenever without use of clauses 8 and 9 we can establish that no matter what truth-values an interpretation may give to certain sentential parts of a sentence ϕ, it will give the value T to ϕ, we can construct a parallel argument for some SC sentence ψ of which ϕ is a substitution instance.These remarks are of course not a *proof* of II, but are designed to make it seem plausible.

Principle II gives us the possibility of establishing the tautologousness of a given sentence by producing a tautologous SC sentence of which it is a substitution instance. In this connection, therefore, it is particularly advantageous to have a way of deciding the tautologousness of SC sentences.

The procedure now to be described, called the *truth-table method,* is sometimes attributed to the philosopher Ludwig Wittgenstein (1889–1951) but in reality is approximately two millennia older than that. It rests upon the following considerations. The only non-logical constants that occur in SC sentences are sentential letters; thus, every SC sentence is built up from sentential letters by successive application of the operations described in clauses (i) and (ii) on page 45. If, bearing this in mind, we look again at the definition of 'normal assignment' we see that the truth-value that a normal assignment gives to an SC sentence is completely determined by the truth-values it gives to the sentential letters occurring therein.

Now, suppose that an SC sentence ϕ contains n distinct sentential letters, where $n \geq 1$. There are only 2^n different ways in which the truth-values T and F can be assigned to these n letters. (Either T or F may be assigned to the first letter; for each of these two $(=2^1)$ choices there are two ways of assigning a truth-value to the second letter, and thus four $(=2^2)$ ways in which T or F may be assigned to the first two letters; for each of these four ways there are two possibilities for assigning a truth-value to the third letter, thus eight ($=2^3$) ways in which the assignment can be made to the first three letters, and so on.) If we consider these 2^n assignments successively, computing in each case the truth-value that would be given to ϕ by any normal assignment which assigned truth-values to the sentential letters of ϕ in just that way, we shall at length establish either that every normal assignment gives the value T to ϕ, or that some normal assignment gives the value F to ϕ.

In testing SC sentences by the above procedure it is convenient to summarize the relevant information by means of so-called *truth-tables.* These are best explained with the help of examples:

P,Q	$((P \rightarrow Q) \vee (Q \rightarrow P))$		
T T	T	T	T
T F	F	T	T
F T	T	T	F
F F	T	T	T

(FIGURE I)

P,Q	$((P \rightarrow Q) \rightarrow (Q \rightarrow P))$		
T T	T	T	T
T F	F	T	T
F T	T	F	F
F F	T	T	T

(FIGURE II)

Figure I is a truth-table for the SC sentence '$((P \rightarrow Q) \vee (Q \rightarrow P))$'; it shows that this sentence is tautologous. More specifically, it is an abbreviation of the following paragraph:

1) Any normal assignment that gives the value T to 'P' and to 'Q' will give T to '$(P \rightarrow Q)$', T to '$(Q \rightarrow P)$', and T to '$((P \rightarrow Q) \vee (Q \rightarrow P))$'.
2) Any normal assignment that gives the value T to 'P' and F to 'Q' will give F to '$(P \rightarrow Q)$', T to '$(Q \rightarrow P)$', and T to '$((P \rightarrow Q) \vee (Q \rightarrow P))$'.
3) Any normal assignment that gives the value F to 'P' and T to 'Q' will give T to '$(P \rightarrow Q)$', F to '$(Q \rightarrow P)$', and T to '$((P \rightarrow Q) \vee (Q \rightarrow P))$'.
4) Any normal assignment that gives the value F to 'P' and to 'Q' will give T to '$(P \rightarrow Q)$', T to '$(Q \rightarrow P)$', and T to '$((P \rightarrow Q) \vee (Q \rightarrow P))$'.

(Since every normal assignment comes under one of these four clauses, it is apparent that '$((P \rightarrow Q) \vee (Q \rightarrow P))$' is assigned the value T by every normal assignment; hence it is tautologous.)

Likewise, the truth-table of Figure II presents the following information in schematic form:

1) Any normal assignment that gives the value T to 'P' and to 'Q' will give T to '$(P \rightarrow Q)$', T to '$(Q \rightarrow P)$', and T to '$((P \rightarrow Q) \rightarrow (Q \rightarrow P))$'.
2) Any normal assignment that gives the value T to 'P' and F to 'Q' will give F to '$(P \rightarrow Q)$', T to '$(Q \rightarrow P)$', and T to '$((P \rightarrow Q) \rightarrow (Q \rightarrow P))$'.
   Etc.

(In view of the third row of this truth-table, we see that the sentence is not tautologous, since there are normal assignments that assign F to 'P' and T to 'Q', resulting in F for '$((P \rightarrow Q) \rightarrow (Q \rightarrow P))$'.)

For an SC sentence containing three sentential letters there will be 2^3, or 8, possible cases to consider. For example, the truth-table for the sentence '$((P \to Q) \lor (Q \to R))$' is:

P,Q,R	$((P \to Q)$	$\lor$	$(Q \to R))$
T T T	T	T	T
T T F	T	T	F
T F T	F	T	T
T F F	F	T	T
F T T	T	T	T
F T F	T	T	F
F F T	T	T	T
F F F	T	T	T

(FIGURE III)

This table says that

1) Any normal assignment that gives the value T to 'P', T to 'Q', and T to 'R' will give T to '$(P \to Q)$', T to '$(Q \to R)$', and T to '$((P \to Q) \lor (Q \to R))$'.
2) Any normal assignment that gives the value T to 'P', T to 'Q', and F to 'R' will give T to '$(P \to Q)$', F to '$(Q \to R)$', and T to '$((P \to Q) \lor (Q \to R))$'. Etc.

(In this case, since the column of entries under the only occurrence of the connective '$\lor$' consists exclusively of 'T', the sentence is tautologous.)

Thus, to construct a truth-table for an SC sentence ϕ, proceed as follows:

1) Write ϕ at the top of the table; at the left, list all the sentential letters that occur in ϕ.

2) Under the list of sentential letters enter on successive lines all possible combinations of truth-values that may be assigned to those letters.

3) Fill out each line according to the definition of 'normal assignment'.

Examples. Theorems 1-100 of the deductive system presented in section 6 below will serve as examples of tautologous SC sentences. In any doubtful case this should be checked by a truth-table. Thus, for theorem 91 we have the table:

P,Q,R	$((P \to Q)$	$\&$	$(Q \to R))$	$\lor$	$(R \to P)$
T T T	T	T	T	T	T
T T F	T	F	F	T	T
T F T	F	F	T	T	T
T F F	F	F	T	T	T
F T T	T	T	T	T	F
F T F	T	F	F	T	T
F F T	T	T	T	T	F
F F F	T	T	T	T	T

(FIGURE IV)

Often, as in this case, the drudgery of writing out a truth-table can be avoided by an analysis in which one works out the consequences of assuming that the given sentence is assigned the value F by some normal assignment. Thus, in order for '$((P \rightarrow Q) \& (Q \rightarrow R)) \vee (R \rightarrow P)$' to have the value F, it is necessary that *both* disjuncts have the value F; but if '$(R \rightarrow P)$' has the value F, then 'R' has the value T and 'P' has F; this in turn ensures that both '$(P \rightarrow Q)$' and '$(Q \rightarrow R)$' have the value T, so that the first disjunct will have the value T. Thus it is impossible for '$((P \rightarrow Q) \& (Q \rightarrow R)) \vee (R \rightarrow P)$' to have the value F.

Further examples of tautologous sentences are obtained by applying principle II to the examples already at hand. Thus, not only is theorem 89 itself tautologous, but so are all of its substitution instances. These will include other SC sentences as well as sentences that contain quantifiers and predicates of arbitrary degree. Here are a few:

$$((Q \rightarrow P) \& (P \rightarrow R)) \vee (R \rightarrow Q)$$
$$((P \rightarrow P) \& (P \rightarrow P)) \vee (P \rightarrow P)$$
$$((P \rightarrow (x)Fx) \& ((x)Fx \rightarrow Q)) \vee (Q \rightarrow P)$$
$$(((x)Fx \rightarrow Fa) \& (Fa \rightarrow (\exists y)Fy)) \vee ((\exists y)Fy \rightarrow (x)Fx).$$

4. *Deciding whether sentences are tautologous.* Although principle II is useful for establishing that sentences are tautologous, it does not, as it stands, provide us with a step-by-step procedure which, applied to an arbitrarily chosen sentence ϕ, will always allow us to decide whether or not ϕ is tautologous. It says, in effect: look and see whether ϕ is a substitution instance of a tautologous SC sentence; if it is, it is tautologous; if not, not. But the trouble is that although if we do find a tautologous SC sentence of which ϕ is a substitution instance we know that ϕ is tautologous, if we do not find such a tautologous SC sentence we can conclude only that either ϕ is not tautologous or we have not looked far enough. What we would like to have is a procedure that will always give us a yes-or-no answer in a finite number of steps.

Economical formulation of such a procedure requires two auxiliary notions:

A sentence ϕ is a *basic truth-functional component* of a sentence ψ if and only if ϕ is atomic or general and occurs free in ψ at least once.

As will be clear upon reflection, the basic truth-functional components of a sentence ϕ are the smallest sentential parts from occurrences of which ϕ may be built up without further use of quantification. This does not preclude that one basic truth-functional component of ϕ may be a proper subsentence of another, as is evident in the two examples given below.

The other auxiliary notion is this: an SC sentence ϕ is *associated with* a sentence ψ if ϕ is obtained from ψ by putting occurrences of sentential letters for all free occurrences in ψ of its basic truth-functional components, distinct components to be replaced by distinct letters, and all free occurrences of the same component to be replaced by occurrences of the same letter.

In terms of these notions, we can now add to I and II in section 3 the following principle:

III. For any sentences ϕ, ψ, if ψ is an SC sentence associated with ϕ, then ϕ is tautologous if and only if ψ is tautologous.

Since for any sentence ϕ we can construct an associated SC sentence ψ, this principle gives a procedure for deciding whether the given sentence ϕ is tautologous or not:

1) Construct an SC sentence ψ associated with ϕ.
2) By a truth-table, test ψ for tautologousness.
3) Then, by III, decide whether ϕ is tautologous.

Example 1. Test '$Fa \rightarrow (x)(Fx \rightarrow Fa)$' for tautologousness. The basic truth-functional components of this sentence are 'Fa' and '$(x)(Fx \rightarrow Fa)$'. Replacing each free occurrence of 'Fa' by an occurrence of 'P' and each free occurrence of '$(x)(Fx \rightarrow Fa)$' by an occurrence of 'Q', we get

$$P \rightarrow Q.$$

But a truth-table test shows that this SC sentence is not tautologous. Hence the original sentence is not tautologous.

Example 2. Test '$Fa \rightarrow ((x)Fx \rightarrow Fa)$' for tautologousness. The basic truth-functional components of this sentence are 'Fa' and '$(x)Fx$'. Replacing each free occurrence of 'Fa' by an occurrence of 'P' and each free occurrence of '$(x)Fx$' by an occurrence of 'Q', we get

$$P \rightarrow (Q \rightarrow P).$$

By the truth-table test we find that this SC sentence is tautologous. Therefore, the original sentence is tautologous.

5. *Further properties.* Each of the following properties of tautologousness or of tautological consequence may easily be established on the basis of the definitions given above. Let ϕ, ψ, χ, θ be any sentences of $\mathfrak{L}$. Then

1) If ϕ is atomic or general, then neither ϕ nor $-\phi$ is tautologous.
2) If $(\phi \rightarrow \psi)$ and ϕ are tautologous, then ψ is tautologous.
3) If $(\phi \rightarrow \psi)$ and $(\psi \rightarrow \chi)$ are tautologous, then $(\phi \rightarrow \chi)$ is tautologous.
4) If ϕ is tautologous, then $(\phi \vee \psi)$, $(\psi \vee \phi)$, $(\psi \rightarrow \phi)$, are tautologous; and if ϕ and ψ are both tautologous, then $(\phi \;\&\; \psi)$, $(\psi \;\&\; \phi)$, $(\phi \leftrightarrow \psi)$, $(\psi \leftrightarrow \phi)$ are tautologous.
5) $(\phi \rightarrow \psi)$ and $(\psi \rightarrow \phi)$ are tautologous if and only if $(\phi \leftrightarrow \psi)$ is tautologous.
6) If $(\phi \leftrightarrow \psi)$ is tautologous, and χ is obtained from θ by replacing one

or more free occurrences of ϕ by occurrences of ψ, then $(\chi \leftrightarrow \theta)$ is tautologous, and χ is tautologous if and only if θ is tautologous.

7) If ϕ is tautologous, and χ is obtained from θ by putting ψ in place of one or more free occurrences of $(\phi \rightarrow \psi)$ or $(\phi \leftrightarrow \psi)$ or $(\psi \leftrightarrow \phi)$ or $(\phi \,\&\, \psi)$ or $(\psi \,\&\, \phi)$, then $(\chi \leftrightarrow \theta)$ is tautologous, and χ is tautologous if and only if θ is tautologous.

8) If ϕ is tautologous, and χ is a basic truth-functional component of ϕ, and ψ is obtained by putting θ in place of each free occurrence of χ in ϕ, then ψ is tautologous.

9) If ϕ is a tautological consequence of the set of sentences Γ and each sentence in Γ is a tautological consequence of the set of sentences Δ, then ϕ is a tautological consequence of Δ.

10) ϕ is a tautological consequence of the set of sentences Γ together with a sentence ψ if and only if $(\psi \rightarrow \phi)$ is a tautological consequence of Γ alone.

6. *Rules of derivation for SC sentences.* In section 4 we described a procedure for deciding whether a given sentence is a tautology. It consists in finding an associated SC sentence, which then is subjected to the truth-table test. This procedure will always work, but when the associated SC sentence contains more than three or four sentential letters the truth-table test becomes quite burdensome (since, as the student will recall, there are 2^n lines in the truth-table for a sentence containing n letters). Hence there is point in looking for more practical ways of showing that SC sentences are tautologous, or, what comes to the same thing, of showing that a given SC sentence is a tautological consequence of a given set of SC sentences.

Suppose, for instance, that the question is raised whether the sentence '$-T$' is a tautological consequence of the set consisting of the following four sentences:

$$P \rightarrow -R$$
$$(S \,\&\, T) \rightarrow R$$
$$-S \rightarrow Q$$
$$-(P \rightarrow Q).$$

If we do a truth-table on the corresponding conditional, which is

$$(((P \rightarrow -R) \,\&\, ((S \,\&\, T) \rightarrow R)) \,\&\, ((-S \rightarrow Q) \,\&\, -(P \rightarrow Q))) \rightarrow -T,$$

we have to write out a table containing 32 lines with 18 entries on each line. A much less onerous method is to try to 'deduce' or 'derive' the formula '$-T$' from the others by making a succession or chain of relatively simple tautological inferences. We might construct some such 'argument' as the following:

(1)	$P \rightarrow -R$	premises
(2)	$(S \& T) \rightarrow R$	
(3)	$-S \rightarrow Q$	
(4)	$-(P \rightarrow Q)$	
(5)	$P \& -Q$	from (4)
(6)	P	from (5)
(7)	$-R$	from (1) and (6)
(8)	$-(S \& T)$	from (2) and (7)
(9)	$S \rightarrow -T$	from (8)
(10)	$-Q$	from (5)
(11)	S	from (3) and (10)
(12)	$-T$	from (9) and (11)

Note that the sentence on each line is a tautological consequence of the sentence or sentences on the lines cited, and that since tautological consequences of tautological consequences of a set Γ are again tautological consequences of Γ, we have found that '$-T$' is a tautological consequence of the given set of premises.

To systematize this sort of procedure we need a reasonably small group of simple inference-rules which together have the property that in general an SC sentence ϕ will be a consequence of a set of SC sentences Γ if and only if there is a correct derivation of ϕ from the set Γ, where a 'correct' derivation is one in which each step is made in accord with one of the rules. It turns out that there are numerous ways of picking such a group. One possibility is presented below. Following our usual practice we first state for reference the essential definitions in a relatively precise and succinct manner, and then offer comments and illustrative examples.

An *SC derivation* is a finite sequence of consecutively numbered lines, each consisting of an SC sentence together with a list of numbers (called the *premise-numbers* of the line), the sequence being constructed according to the following rules (letting ϕ, ψ, χ, θ be any SC sentences):

P (Introduction of premises) Any SC sentence may be entered on a line, with the line number taken as the only premise-number.

MP (Modus ponens) ψ may be entered on a line if ϕ and $(\phi \rightarrow \psi)$ appear on earlier lines; as premise-numbers of the new line take all premise-numbers of those earlier lines.

MT (Modus tollens) ϕ may be entered on a line if ψ and $(-\phi \rightarrow -\psi)$ appear on earlier lines; as premise-numbers of the new line take all premise-numbers of those earlier lines.

C (Conditionalization) $(\phi \rightarrow \psi)$ may be entered on a line if ψ appears on an earlier line; as premise-numbers of the new line take all those of the earlier line, with the exception (if desired) of any that is the line number of a line on which ϕ appears.

D (Definitional interchange) If ψ is obtained from ϕ by replacing an occurrence of a sentence χ in ϕ by an occurrence of a sentence to which χ is definitionally equivalent (see below), then ψ may be entered on a line if ϕ appears on an earlier line; as premise-numbers of the new line take those of the earlier line.

For any SC sentences ϕ, ψ:

$(\phi \vee \psi)$ is *definitionally equivalent to* $(-\phi \rightarrow \psi)$, and vice versa;
$(\phi \,\&\, \psi)$ is *definitionally equivalent to* $-(\phi \rightarrow -\psi)$, and vice versa;
$(\phi \leftrightarrow \psi)$ is *definitionally equivalent to* $((\phi \rightarrow \psi) \,\&\, (\psi \rightarrow \phi))$, and vice versa.

An SC derivation in which an SC sentence ϕ appears on the last line and in which all premises of that line* belong to a set of SC sentences Γ, is called an *SC derivation* (or *proof*) *of* ϕ *from* Γ.

An SC sentence ϕ is *SC derivable* from a set of SC sentences Γ if and only if there is an SC derivation of ϕ from Γ.

An SC sentence ϕ is an *SC theorem* if and only if ϕ is SC derivable from Λ (the empty set of sentences).

To illustrate the application of these rules we now give SC derivations for a number of SC theorems. Where a theorem is listed without proof it is expected that the student will demonstrate it for himself. He should also endeavor to find different and, if possible, more elegant ways of deriving the theorems for which derivations are given. The gain in all this will be twofold: the student will become familiar with the idea of a formal derivation, and he will get a working acquaintance with a large variety of tautologous SC sentences. He will thus be provided with an introduction to the more complex derivations to be studied in the next chapter, as well as with a stock of 'ammunition' for the efficacious employment of the rules there set forth.

1. $(P \rightarrow Q) \rightarrow ((Q \rightarrow R) \rightarrow (P \rightarrow R))$ (Principle of the syllogism)

{1}	(1) $P \rightarrow Q$	P
{2}	(2) $Q \rightarrow R$	P
{3}	(3) P	P
{1,3}	(4) Q	(1)(3) MP
{1,2,3}	(5) R	(2)(4) MP
{1,2}	(6) $P \rightarrow R$	(3)(5) C
{1}	(7) $(Q \rightarrow R) \rightarrow (P \rightarrow R)$	(2)(6) C
Λ	(8) $(P \rightarrow Q) \rightarrow ((Q \rightarrow R) \rightarrow (P \rightarrow R))$	(1)(7) C

*To say that a sentence ϕ is a premise of a given line is to say that ϕ appears on a line numbered by a premise-number of the given line; and to say that a sentence ϕ appears on a line is to say that the line consists of ϕ and a set of numbers.

Comments: The foregoing eight lines constitute an SC derivation of the SC sentence ‘$(P \rightarrow Q) \rightarrow ((Q \rightarrow R) \rightarrow (P \rightarrow R))$’ from the empty set of sentences. The premise-numbers of each line are indicated by the numerals appearing in braces; the line number is given by the numeral in parentheses. Off to the right we place some annotations to help the reader check that the derivation is constructed in accord with the rules. The first three lines have been entered by rule P; in each case the line number has been taken as the only premise-number. The sentence on the fourth line has been obtained by *modus ponens* from those appearing on lines (1) and (3); as premise-numbers of the fourth line we have taken those of lines (1) and (3). Likewise, the fifth line has been obtained on the basis of rule MP from lines (2) and (4); as premise-numbers we have taken those of lines (2) and (4). This does not mean that we have taken 2 and 4 as premise-numbers of line (5), but rather that we have taken 1, 2 and 3, which are all the premise-numbers of lines (2) and (4). Line (6) has been obtained from lines (3) and (5) by rule C. As permitted by that rule we have dropped 3 from the set of premise-numbers. The intuitive basis for this is that if ‘R’ is a tautological consequence of the sentences ‘$P \rightarrow Q$’, ‘$Q \rightarrow R$’, and ‘P’ (which, in effect, is what line (5) says), then ‘$P \rightarrow R$’ is a tautological consequence of ‘$P \rightarrow Q$’ and ‘$Q \rightarrow R$’ alone. Lines (7) and (8) are obtained by further application of rule C. Each time rule C is applied we drop a premise, until, in line (8), we have none left. Thus the sentence appearing on line (8) is a tautological consequence of the empty set of sentences; i.e., it is a tautology.

2. $(Q \rightarrow R) \rightarrow ((P \rightarrow Q) \rightarrow (P \rightarrow R))$
3. $P \rightarrow ((P \rightarrow Q) \rightarrow Q)$

{1}	(1) P	P
{2}	(2) $P \rightarrow Q$	P
{1,2}	(3) Q	(1)(2) MP
{1}	(4) $(P \rightarrow Q) \rightarrow Q$	(2)(3) C
Λ	(5) $P \rightarrow ((P \rightarrow Q) \rightarrow Q)$	(1)(4) C

4. $(P \rightarrow (Q \rightarrow R)) \rightarrow ((P \rightarrow Q) \rightarrow (P \rightarrow R))$
5. $(P \rightarrow (Q \rightarrow R)) \rightarrow (Q \rightarrow (P \rightarrow R))$
6. $P \rightarrow P$ (Law of identity)

{1}	(1) P	P
Λ	(2) $P \rightarrow P$	(1) C

Comment: In the foregoing two-step derivation, rule C has been applied in a rather unusual way. To see that this application is legitimate, refer to the rule and substitute ‘‘P’’ for both ‘ϕ’ and ‘ψ’.

7. $Q \rightarrow (P \rightarrow Q)$

{1}	(1) Q	P
{1}	(2) $P \rightarrow Q$	(1) C
Λ	(3) $Q \rightarrow (P \rightarrow Q)$	(1)(2) C

Comment: This case illustrates another unusual application of rule C. In inferring line (2) we are unable to drop any premises, for 'P' is not a premise of line (1).

Henceforward we shall often omit the annotations, assuming that the reader will have no difficulty in supplying them for himself.

8. $-P \to (P \to Q)$ (Law of Duns Scotus)

{1}	(1) $-P$	
{2}	(2) P	
{1}	(3) $-Q \to -P$	(1) C
{1,2}	(4) Q	(2)(3) MT
{1}	(5) $P \to Q$	
Λ	(6) $-P \to (P \to Q)$	

9. $P \to (-P \to Q)$

10. $--P \to P$

{1}	(1) $--P$	
{2}	(2) $-P$	
{1}	(3) $----P \to --P$	(1) C
{1,2}	(4) $---P$	(2)(3) MT
{1}	(5) $-P \to ---P$	
{1}	(6) P	(1)(5) MT
Λ	(7) $--P \to P$	

At this point it is advantageous to give explicit consideration to a certain matter that may already have occurred to the student: if we have a derivation of an SC sentence ϕ from the empty set, we can easily construct a similar derivation for any SC sentence ψ that is a substitution instance of ϕ. We need only go through our given derivation of ϕ, making everywhere the same substitutions as are required to transform ϕ into ψ, and the result will be a derivation of ψ.

Thus, corresponding to our derivation of theorem 6 we have

{1}	(1) Q	P
Λ	(2) $Q \to Q$	(1) C

as a derivation for '$Q \to Q$', and

{1}	(1) $R \leftrightarrow -S$	P
Λ	(2) $(R \leftrightarrow -S) \to (R \leftrightarrow -S)$	(1) C

as a derivation for '$(R \leftrightarrow -S) \to (R \leftrightarrow -S)$', and, in general,

{1}	(1) ψ	P
Λ	(2) $\psi \to \psi$	(1) C

as a derivation for $\psi \to \psi$.

Under the kind of substitution described, a premise (entered by rule P) goes over into a premise, a *modus ponens* argument into a *modus ponens* argument, a *modus tollens* argument into a *modus tollens* argument, and similarly for rules C and D. Thus a correct derivation goes over into a correct derivation, and a correct derivation from the empty set goes over into a correct derivation from the empty set. Consequently, if an SC sentence ϕ is an SC theorem and an SC sentence ψ is a substitution instance of ϕ, then ψ is an SC theorem.

This fact is useful in connection with a scheme for abbreviating derivations. In constructing derivations we often find that it would be very convenient to be able to use a previously proved theorem, or a substitution instance of a previously proved theorem. Now, of course, this can be done by interpolating, at the point where the theorem is needed, a proof of that theorem; the drawbacks to such a procedure are that it would require us to repeat the same proofs over and over, and that it would make proofs exceedingly long (for we may wish to use previously proved theorems that were proved with the help of other previously proved theorems, and so on). We are thus motivated to formulate the following 'short-cut' rule of inference:

TH Any SC sentence that is a substitution instance of a previously proved SC theorem may be entered on a line, with the empty set of premise-numbers; more generally, ψ may be entered on a line if $\phi_1, \phi_2, \ldots, \phi_n$ appear on earlier lines and the conditional

$$(\phi_1 \rightarrow (\phi_2 \rightarrow \ldots \rightarrow (\phi_n \rightarrow \psi) \ldots))$$

is a substitution instance of an already proved SC theorem; as premise-numbers of the new line take all premise-numbers of those earlier lines.

Rule TH, examples of the application of which will be found in the proofs of theorems 11 and 13 below, is not on a par with the basic rules P, MP, MT, C, and D, for any inference that can be made with its help can also be made by using the basic rules alone. Where, as in the proof of theorem 11 below, we utilize TH to introduce a substitution instance of a previously proved theorem, we could simply have inserted a proof of that substitution instance. For theorem 11 this would give the following proof in which TH is not used:

{1}	(1)	$---P$
{2}	(2)	$--P$
{1}	(3)	$-----P \rightarrow ---P$
{1,2}	(4)	$----P$
{1}	(5)	$--P \rightarrow -----P$
{1}	(6)	$-P$
Λ	(7)	$---P \rightarrow -P$

{8}	(8) P	
{8}	(9) $--P$	
Λ	(10) $P \rightarrow --P$	

The first seven lines of this proof are simply our derivation of theorem 10 (with '$-P$' for 'P', so as to obtain the substitution instance '$---P \rightarrow -P$' instead of the theorem itself). The last four lines are our derivation of theorem 11, with premise-numbers increased by 6.

Similarly when, as in the fourth line of the derivation of theorem 13, we use TH to make an inference for which the corresponding conditional is an already proved theorem, we could have inserted a proof of that conditional and then detached our conclusion by using MP. In general, if sentences $\phi_1, \phi_2, \ldots, \phi_n$ appear on earlier lines and the conditional

$$(\phi_1 \rightarrow (\phi_2 \rightarrow \ldots \rightarrow (\phi_n \rightarrow \psi)\,..))$$

is a substitution instance of an already proved theorem, we can obtain ψ without using TH if we insert a proof of the conditional and then apply MP n times to detach ψ.

Therefore, TH is only a device for abbreviating proofs; it adds convenience, but not strength to our deductive system.

11. $P \rightarrow --P$

Λ	(1) $---P \rightarrow -P$	TH 10
{2}	(2) P	P
{2}	(3) $--P$	(1)(2) MT
Λ	(4) $P \rightarrow --P$	(2)(3) C

12. $(-P \rightarrow -Q) \rightarrow (Q \rightarrow P)$

{1}	(1) $(-P \rightarrow -Q)$	P
{2}	(2) Q	P
{1,2}	(3) P	(1)(2) MT
{1}	(4) $Q \rightarrow P$	(2)(3) C
Λ	(5) $(-P \rightarrow -Q) \rightarrow (Q \rightarrow P)$	(1)(4) C

13. $(P \rightarrow -Q) \rightarrow (Q \rightarrow -P)$

{1}	(1) $P \rightarrow -Q$	P
{2}	(2) Q	P
{3}	(3) $--P$	P
{3}	(4) P	(3) TH 10
{1,3}	(5) $-Q$	(1)(4) MP
{1}	(6) $--P \rightarrow -Q$	(3)(5) C
{1,2}	(7) $-P$	(2)(6) MT
{1}	(8) $Q \rightarrow -P$	(2)(7) C
Λ	(9) $(P \rightarrow -Q) \rightarrow (Q \rightarrow -P)$	(1)(8) C

14. $(-P \rightarrow Q) \rightarrow (-Q \rightarrow P)$

15. $(P \to Q) \to (-Q \to -P)$ (Principle of transposition)

$\{1\}$	(1) $P \to Q$	P
Λ	(2) $Q \to --Q$	TH 11
$\{1\}$	(3) $P \to --Q$	(1)(2) TH 1
$\{1\}$	(4) $-Q \to -P$	(3) TH 13
Λ	(5) $(P \to Q) \to (-Q \to -P)$	(1)(4) C

16. $(-P \to P) \to P$ (Law of Clavius)

$\{1\}$	(1) $(-P \to P)$	
Λ	(2) $-P \to (P \to -(-P \to P))$	TH 8
Λ	(3) $(-P \to P) \to (-P \to -(-P \to P))$	(2) TH 4
$\{1\}$	(4) $-P \to -(-P \to P)$	
$\{1\}$	(5) $(-P \to P) \to P$	(4) TH 12
$\{1\}$	(6) P	
Λ	(7) $(-P \to P) \to P$	

17. $(P \to -P) \to -P$

18. $-(P \to Q) \to P$

Λ	(1) $-P \to (P \to Q)$	TH 8
Λ	(2) $-(P \to Q) \to P$	(1) TH 14

19. $-(P \to Q) \to -Q$

20. $P \to (Q \to (P \& Q))$

$\{1\}$	(1) P	
$\{2\}$	(2) $--(P \to -Q)$	
$\{2\}$	(3) $P \to -Q$	(2) TH 10
$\{1,2\}$	(4) $-Q$	(1)(3) MP
$\{1\}$	(5) $--(P \to -Q) \to -Q$	
$\{1\}$	(6) $Q \to -(P \to -Q)$	(5) TH 12
$\{1\}$	(7) $Q \to (P \& Q)$	(6) D
Λ	(8) $P \to (Q \to (P \& Q))$	

21. $(P \to Q) \to ((Q \to P) \to (P \leftrightarrow Q))$

22. $(P \leftrightarrow Q) \to (P \to Q)$

23. $(P \leftrightarrow Q) \to (Q \to P)$

24. $(P \vee Q) \leftrightarrow (Q \vee P)$ (Commutative law for disjunction)

$\{1\}$	(1) $P \vee Q$	
$\{1\}$	(2) $-P \to Q$	(1) D
$\{1\}$	(3) $-Q \to P$	(2) TH 14
$\{1\}$	(4) $Q \vee P$	
Λ	(5) $(P \vee Q) \to (Q \vee P)$	
$\{6\}$	(6) $Q \vee P$	
$\{6\}$	(7) $-Q \to P$	
$\{6\}$	(8) $-P \to Q$	(7) TH 14
$\{6\}$	(9) $P \vee Q$	
Λ	(10) $(Q \vee P) \to (P \vee Q)$	
Λ	(11) $(P \vee Q) \leftrightarrow (Q \vee P)$	(5)(10) TH 21

25. $P \rightarrow (P \vee Q)$
26. $Q \rightarrow (P \vee Q)$
27. $(P \vee P) \leftrightarrow P$ (Principle of tautology for disjunction)

{1}	(1) $P \vee P$	
{1}	(2) $-P \rightarrow P$	
{1}	(3) P	(2) TH 16
Λ	(4) $(P \vee P) \rightarrow P$	
Λ	(5) $P \rightarrow (P \vee P)$	TH 25
Λ	(6) $(P \vee P) \leftrightarrow P$	(4)(5) TH 21

28. $P \leftrightarrow P$
29. $--P \leftrightarrow P$ (Principle of double negation)
30. $(P \leftrightarrow Q) \leftrightarrow (Q \leftrightarrow P)$
31. $(P \leftrightarrow Q) \leftrightarrow (-P \leftrightarrow -Q)$
32. $(P \leftrightarrow Q) \rightarrow ((P \,\&\, R) \leftrightarrow (Q \,\&\, R))$
33. $(P \leftrightarrow Q) \rightarrow ((R \,\&\, P) \leftrightarrow (R \,\&\, Q))$
34. $(P \leftrightarrow Q) \rightarrow ((P \vee R) \leftrightarrow (Q \vee R))$
35. $(P \leftrightarrow Q) \rightarrow ((R \vee P) \leftrightarrow (R \vee Q))$
36. $(P \leftrightarrow Q) \rightarrow ((P \rightarrow R) \leftrightarrow (Q \rightarrow R))$
37. $(P \leftrightarrow Q) \rightarrow ((R \rightarrow P) \leftrightarrow (R \rightarrow Q))$
38. $(P \leftrightarrow Q) \rightarrow ((P \leftrightarrow R) \leftrightarrow (Q \leftrightarrow R))$
39. $(P \leftrightarrow Q) \rightarrow ((R \leftrightarrow P) \leftrightarrow (R \leftrightarrow Q))$

Now suppose that ϕ, ψ, χ, θ, and θ_1 are any SC sentences. By virtue of theorem 31 we see that if $\chi \leftrightarrow \theta$ is a theorem, then $-\chi \leftrightarrow -\theta$ is a theorem. Similarly, from theorem 32 we see that if $\chi \leftrightarrow \theta$ is a theorem, then $(\chi \,\&\, \theta_1) \leftrightarrow (\theta \,\&\, \theta_1)$ is a theorem. Obviously we could also apply both 31 and 32 in a single case, to obtain such a result as: if $\chi \leftrightarrow \theta$ is a theorem, then

$$-(-\chi \,\&\, \theta_1) \leftrightarrow -(-\theta \,\&\, \theta_1)$$

is a theorem. In general, by virtue of theorems 31 through 39, we see that if $\chi \leftrightarrow \theta$ is a theorem and if ϕ is obtained from ψ by replacing an occurrence of θ in ψ by an occurrence of χ then $\phi \leftrightarrow \psi$ is a theorem. For ϕ can be built up from that occurrence of χ by means of the connectives, and ψ can be built up in an exactly corresponding way from the occurrence of θ; hence, by (possibly) repeated application of (some or all of) theorems 31–39, we have the result that $\phi \leftrightarrow \psi$ is a theorem. On the basis of this, together with theorem 22 and rule TH, the following very useful 'short-cut' rule is justified.

R (Replacement) ϕ may be entered on a line if ψ appears on an earlier line and ϕ is obtained from ψ by replacing an occurrence of θ in ψ by an occurrence of χ, provided that $\chi \leftrightarrow \theta$ or $\theta \leftrightarrow \chi$ is a substitution instance of a previously proved SC theorem; as premise numbers of the new line take those of the earlier line.

The application of rule R is illustrated in the proof of theorem 41 below.

40. $(P \vee (Q \vee R)) \leftrightarrow (Q \vee (P \vee R))$
41. $(P \vee (Q \vee R)) \leftrightarrow ((P \vee Q) \vee R)$ (Associative law for disjunction)

Λ	(1) $(P \vee (Q \vee R)) \leftrightarrow (Q \vee (P \vee R))$	TH 40
Λ	(2) $(P \vee (Q \vee R)) \leftrightarrow (Q \vee (R \vee P))$	(1) 24 R
Λ	(3) $(P \vee (Q \vee R)) \leftrightarrow (R \vee (Q \vee P))$	(2) 40 R
Λ	(4) $(P \vee (Q \vee R)) \leftrightarrow ((Q \vee P) \vee R)$	(3) 24 R
Λ	(5) $(P \vee (Q \vee R)) \leftrightarrow ((P \vee Q) \vee R)$	(4) 24 R

42. $-(P \& Q) \leftrightarrow (-P \vee -Q)$
43. $-(P \vee Q) \leftrightarrow (-P \& -Q)$
44. $(P \& Q) \leftrightarrow -(-P \vee -Q)$
45. $(P \vee Q) \leftrightarrow -(-P \& -Q)$

(42–45: De Morgan's laws)

46. $(P \& Q) \leftrightarrow (Q \& P)$ (Commutative law for conjunction)
47. $(P \& Q) \rightarrow P$
48. $(P \& Q) \rightarrow Q$

(47–48: Laws of simplification)

49. $(P \& P) \leftrightarrow P$ (Principle of tautology for conjunction)

Λ	(1) $(P \& P) \rightarrow P$	TH 47
Λ	(2) $(P \rightarrow -P) \rightarrow -P$	TH 17
Λ	(3) $P \rightarrow -(P \rightarrow -P)$	(2) TH 13
Λ	(4) $P \rightarrow (P \& P)$	(3) D
Λ	(5) $(P \& P) \leftrightarrow P$	(1)(4) TH 21

50. $(P \& (Q \& R)) \leftrightarrow ((P \& Q) \& R)$ (Associative law for conjunction)
51. $(P \rightarrow (Q \rightarrow R)) \leftrightarrow ((P \& Q) \rightarrow R)$ (Export-import law)
52. $(P \rightarrow Q) \leftrightarrow -(P \& -Q)$
53. $(P \rightarrow Q) \leftrightarrow (-P \vee Q)$
54. $(P \vee (Q \& R)) \leftrightarrow ((P \vee Q) \& (P \vee R))$
55. $(P \& (Q \vee R)) \leftrightarrow ((P \& Q) \vee (P \& R))$

(54–55: Distributive laws)

56. $((P \& Q) \vee (R \& S)) \leftrightarrow (((P \vee R) \& (P \vee S)) \& ((Q \vee R) \& (Q \vee S)))$
57. $P \rightarrow ((P \& Q) \leftrightarrow Q)$
58. $P \rightarrow ((Q \& P) \leftrightarrow Q)$
59. $P \rightarrow ((P \rightarrow Q) \leftrightarrow Q)$
60. $P \rightarrow ((P \leftrightarrow Q) \leftrightarrow Q)$
61. $P \rightarrow ((Q \leftrightarrow P) \leftrightarrow Q)$
62. $-P \rightarrow ((P \vee Q) \leftrightarrow Q)$
63. $-P \rightarrow ((Q \vee P) \leftrightarrow Q)$
64. $-P \rightarrow (-(P \leftrightarrow Q) \leftrightarrow Q)$
65. $-P \rightarrow (-(Q \leftrightarrow P) \leftrightarrow Q)$
66. $P \vee -P$ (Law of excluded middle)
67. $-(P \& -P)$ (Law of contradiction)
68. $(P \leftrightarrow Q) \leftrightarrow ((P \& Q) \vee (-P \& -Q))$
69. $-(P \leftrightarrow Q) \leftrightarrow (P \leftrightarrow -Q)$
70. $((P \leftrightarrow Q) \& (Q \leftrightarrow R)) \rightarrow (P \leftrightarrow R)$
71. $((P \leftrightarrow Q) \leftrightarrow P) \leftrightarrow Q$

72. $(P \leftrightarrow (Q \leftrightarrow R)) \leftrightarrow ((P \leftrightarrow Q) \leftrightarrow R)$
73. $(P \rightarrow Q) \leftrightarrow (P \rightarrow (P \& Q))$
74. $(P \rightarrow Q) \leftrightarrow (P \leftrightarrow (P \& Q))$
75. $(P \rightarrow Q) \leftrightarrow ((P \vee Q) \rightarrow Q)$
76. $(P \rightarrow Q) \leftrightarrow ((P \vee Q) \leftrightarrow Q)$
77. $(P \rightarrow Q) \leftrightarrow (P \rightarrow (P \rightarrow Q))$
78. $(P \rightarrow (Q \& R)) \leftrightarrow ((P \rightarrow Q) \& (P \rightarrow R))$
79. $((P \vee Q) \rightarrow R) \leftrightarrow ((P \rightarrow R) \& (Q \rightarrow R))$
80. $(P \rightarrow (Q \vee R)) \leftrightarrow ((P \rightarrow Q) \vee (P \rightarrow R))$
81. $((P \& Q) \rightarrow R) \leftrightarrow ((P \rightarrow R) \vee (Q \rightarrow R))$
82. $(P \rightarrow (Q \leftrightarrow R)) \leftrightarrow ((P \& Q) \leftrightarrow (P \& R))$
83. $((P \& -Q) \rightarrow R) \leftrightarrow (P \rightarrow (Q \vee R))$
84. $(P \vee Q) \leftrightarrow ((P \rightarrow Q) \rightarrow Q)$
85. $(P \& Q) \leftrightarrow ((Q \rightarrow P) \& Q)$
86. $(P \rightarrow Q) \vee (Q \rightarrow R)$
87. $(P \rightarrow Q) \vee (-P \rightarrow Q)$
88. $(P \rightarrow Q) \vee (P \rightarrow -Q)$
89. $((P \& Q) \rightarrow R) \leftrightarrow ((P \& -R) \rightarrow -Q)$
90. $(P \rightarrow Q) \rightarrow ((R \rightarrow (Q \rightarrow S)) \rightarrow (R \rightarrow (P \rightarrow S)))$
91. $((P \rightarrow Q) \& (Q \rightarrow R)) \vee (R \rightarrow P)$
92. $((P \rightarrow Q) \& (R \rightarrow S)) \rightarrow ((P \vee R) \rightarrow (Q \vee S))$
93. $((P \rightarrow Q) \vee (R \rightarrow S)) \leftrightarrow ((P \rightarrow S) \vee (R \rightarrow Q))$
94. $((P \vee Q) \rightarrow R) \leftrightarrow ((P \rightarrow R) \& ((-P \& Q) \rightarrow R))$
95. $((P \rightarrow Q) \rightarrow (Q \rightarrow R)) \leftrightarrow (Q \rightarrow R)$
96. $((P \rightarrow Q) \rightarrow (Q \rightarrow R)) \rightarrow ((P \rightarrow Q) \rightarrow (P \rightarrow R))$
97. $((P \rightarrow Q) \rightarrow R) \rightarrow ((R \rightarrow P) \rightarrow P)$
98. $((P \rightarrow Q) \rightarrow R) \rightarrow ((P \rightarrow R) \rightarrow R)$
99. $(-P \rightarrow R) \rightarrow ((Q \rightarrow R) \rightarrow ((P \rightarrow Q) \rightarrow R))$
100. $(((P \rightarrow Q) \rightarrow R) \rightarrow S) \rightarrow ((Q \rightarrow R) \rightarrow (P \rightarrow S))$

It is not difficult to see that if a sentence ϕ appears on a line of a derivation constructed in accord with the five rules P, MP, MT, C, and D, then ϕ is a tautological consequence of the premises of that line. We show this by establishing that (1) any sentence appearing on the first line of such a derivation is a tautological consequence of the premises of that line, and (2) any sentence appearing on a later line is a tautological consequence of its premises if all sentences appearing on earlier lines are tautological consequences of theirs.

As concerns (1): if ϕ appears on the first line, ϕ was entered by rule P and hence is its own only premise.

To establish (2) we consider the rules one at a time.

(i) If ϕ was entered by rule P, it is obviously a tautological consequence of the premises of its line.

(ii) If ϕ was entered by rule MP, its premises are all those of a pair of earlier lines on which sentences ψ and $(\psi \to \phi)$ appear. Now ϕ is a tautological consequence of ψ and $(\psi \to \phi)$, and since by hypothesis these are tautological consequences of the premises of the earlier lines on which they appear, ϕ is a tautological consequence of those premises too (p. 96, item 9).

(iii) If ϕ was entered by rule MT, the argument is exactly analogous to that for MP.

(iv) If ϕ was entered by rule C, then $\phi = (\psi \to \chi)$, where χ appears on an earlier line. By hypothesis, χ is a tautological consequence of the premises of its line, which may include ψ. Therefore, $(\psi \to \chi)$ is a tautological consequence of these premises excluding ψ (p. 96, items 9 and 10).

(v) Finally, if ϕ was entered by rule D, some sentence from which ϕ may be obtained by a definitional interchange appears on an earlier line. But such sentences are always tautological consequences of one another. (The proof of this is left as an exercise to the reader.) Hence ϕ is a consequence of the premises of that earlier line, and these are none other than the premises of ϕ.

Therefore, any sentence appearing on a line of an SC derivation is a tautological consequence of the premises of that line. In particular, if we are able to derive an SC sentence from the empty set, we thereby establish that it is a tautology. (This does not imply, of course, that if we *fail* to find a derivation we have established that the given sentence is *not* a tautology; for that purpose a truth-table analysis is still required.)

In addition to what has just been shown, our five rules have another important property, namely, that if an SC sentence ϕ is a tautological consequence of a set of SC sentences Γ, then ϕ is derivable from Γ by the rules. The demonstration of this will be given as an exercise later.

EXERCISES

1. Construct a truth-table for each of the following SC sentences.
   (a) $(P \,\&\, Q) \to (P \vee Q)$
   (b) $((P \to Q) \to P) \to Q$
   (c) $((P \leftrightarrow -Q) \leftrightarrow -P) \leftrightarrow Q$
   (d) $Q \leftrightarrow ((P \,\&\, Q) \vee (-P \,\&\, -Q))$
   (e) $((P \to Q) \to P) \leftrightarrow P$ (Peirce's Law)
   (f) $(P \to (Q \,\&\, -Q)) \to -P$
   (g) $(P \to (Q \to R)) \to ((P \to Q) \to R)$
   (h) $P \to ((Q \to R) \to ((P \to Q) \to R))$
   (i) $((P \to Q) \to R) \to ((P \to R) \to R)$
   (j) $(P \to Q) \vee (P \leftrightarrow -Q)$

2. Find an SC sentence containing the letters 'P', 'Q', 'R', and no others, and such that in its truth-table the column of entries under the principal connective-occurrence (when the combinations of truth-values are listed as in Figure III, page 93) is:

T
T
T
F
F
T
T
T

3. List the basic truth-functional components of each of the following sentences.
   (a) $P \rightarrow (Q \vee R)$
   (b) $(x)Fx \rightarrow (Fa \rightarrow R)$
   (c) $(x)(y)Fxy$
   (d) $(x)(Fxa \rightarrow (\exists y)Gxay)$
   (e) $Hab \rightarrow (\exists x)(Hxb \leftrightarrow Hab)$
   (f) $(x)Fxa \leftrightarrow (y)(x)Fxa$
   (g) $(P \,\&\, (x)Fx) \leftrightarrow (-P \,\&\, (x)Gx)$
   (h) $(x)(Fx \,\&\, (P \vee -P))$
   (i) $Q \rightarrow (x)(P \,\&\, Q)$
   (j) $((z)Gz \,\&\, (\exists y)Hy) \leftrightarrow -(z)(\exists y)(Gz \,\&\, Hy)$
4. (a) For each of the sentences of exercise 3 give an associated SC sentence.
   (b) Give an SC sentence of which every sentence is a substitution instance. How many such SC sentences are there?
   (c) Give an SC sentence of which no substitution instance is a tautology.
   (d) Give an SC sentence of which every conditional sentence is a substitution instance.
   (e) What are the basic truth-functional components of an SC sentence?
   (f) Explain why it is always the case that, if an SC sentence ϕ is associated with an SC sentence ψ, then ψ is associated with ϕ.
5. Determine which of the following sentences are tautologous (by constructing for each an associated SC sentence and using truth-tables).
   (a) $(Ca \,\&\, Gm) \rightarrow Ca$
   (b) $-(-Aa \,\&\, (Aa \vee P)) \vee P$
   (c) $(Ha \vee -Dj) \rightarrow (-Ha \rightarrow -Dj)$
   (d) $(Ca \vee (x)(Fx \rightarrow Ca)) \leftrightarrow Ca$
   (e) $(x)(Fx \vee Ga) \rightarrow ((x)Fx \vee Ga)$
   (f) $((-(\exists y)Dy \rightarrow (\exists y)Dy) \vee -Da) \rightarrow (\exists y)Dy$
   (g) $(x)(Fx \vee -Fx) \rightarrow ((x)(Fx \vee -Fx) \vee (x)-(Fx \vee -Fx))$
   (h) $(x)(Fx \,\&\, (y)(Gy \rightarrow (\exists z)Fz)) \rightarrow (x)(Fx \vee (\exists z)Fz)$
   (i) $(((x)Fx \,\&\, (y)Gy) \rightarrow (\exists z)Fz) \rightarrow ((x)Fx \vee (\exists z)Fz)$
   (j) $Q \rightarrow (-(x)(y)Cxy \rightarrow (Q \,\&\, -(x)(y)Cyx))$
6. From the empty set derive each of the tautologies in exercise 1. (Use any or all of the rules given in section 6, and take theorems 1–90 as previously proved.)

7. For each of the following groups derive the last sentence from the remaining sentences. (Use the rules of section 6; take theorems 1-100 as previously proved.)

(a) $P \to Q$
$-P \to R$
$-Q \to -R$
Q

(b) $P \to Q$
$(Q \to P) \to P$
Q

(c) $(P \to Q) \to R$
$S \to -P$
T
$(-S \& T) \to Q$
R

(d) $P \to Q$
$R \to -S$
$S \to P$
$Q \to R$
$S \to T$

(e) $(-P \to Q) \to R$
$-R$
$-Q \to P$
S

(f) $P \to R$
$Q \to -S$
$R \to Q$
$P \to -S$

(g) $(P \& Q) \to R$
$R \to S$
$Q \& -S$
$-P$

(h) $-(P \& -Q) \vee -(-S \& -T)$
$-(T \vee Q)$
$U \to (-T \to (-S \& P))$
$-U$

(i) $(P \& -Q) \vee (P \& R)$
$-Q \to -P$
R

(j) $(P \to Q) \to Q$
$(T \to P) \to R$
$(R \to S) \to -(S \to Q)$
R

(k) $(P \to Q) \vee (R \to S)$
$(P \to S) \vee (R \to Q)$

8. For each of the following arguments try to symbolize premises and conclusion by means of SC sentences and then derive the symbolized conclusion from the symbolized premises. (Since only sentential letters are involved, the interpretation need not be specified.)

(a) Either the goal will be scored or Mulligan will fall flat on his face. If Mulligan falls flat on his face the fans will jeer. The fans will not jeer. Therefore, it is not the case that if the goal is scored the fans will jeer.

(b) The students are happy if and only if no test is given. If the students are happy, the professor feels good. But if the professor feels good, he is in no condition to lecture, and if he is in no condition to lecture a test is given. Therefore, the students are not happy.

(c) If the portrait resembles the customer, then he and the artist will be disappointed. If the portrait does not resemble the customer, his wife will refuse to pay; and, if that happens, the artist will be disappointed. Therefore, the artist will be disappointed.

(d) If Henry gets a Rolls-Royce for Christmas, he will have to park it in the street unless someone gives him a garage, too. It is not true that if he is a good boy then someone will give him a garage. But if Henry is a good boy he will get a Rolls-Royce for Christmas. Therefore, he will indeed get a Rolls-Royce for Christmas and will have to park it in the street.

9. On the basis of the definition of 'tautologous sentence' demonstrate statements 1–3 on page 95.

10. Let ϕ be an SC sentence not containing 'v', '&', or '→'. Show by induction that ϕ is an SC theorem if and only if the following conditions hold jointly:
    1) each sentential letter that occurs in ϕ occurs an even number of times, and
    2) if '−' occurs in ϕ it occurs an even number of times.
11. Treat each of the following arguments as in exercise 8.
    (a) If Moriarty is caught, then London will become a singularly uninteresting city from the criminological point of view. Moriarty will be caught if and only if Holmes escapes him. But in any case Holmes will return. Therefore, if Holmes escapes Moriarty, then London will become a singularly uninteresting city from the criminological point of view and Watson will sell the violin and syringe unless Holmes returns.
    (b) If Holmes is successful and Colonel Moran is apprehended, then the famous air gun of Von Herder will embellish the Scotland Yard Museum. But it is not true that the famous air-gun of Von Herder will embellish the Scotland Yard Museum if Colonel Moran is apprehended. Therefore, if Colonel Moran is apprehended, Holmes will not be successful.
12. Show that, for any SC sentence ϕ not containing the negation sign '−', the result of substituting '$P \to P$' for each occurrence of each sentential letter in ϕ is a tautology.
13. From the empty set derive each of the following. (Use the rules of section 6; take theorems 1–100 as previously proved.)
    (a) $((P \to Q) \to Q) \leftrightarrow ((Q \to P) \to P)$
    (b) $(P \to Q) \to ((-P \to Q) \to Q)$
    (c) $(((P \vee Q) \,\&\, (P \to R)) \,\&\, (Q \to R)) \to R$
    (d) $(P \leftrightarrow Q) \leftrightarrow ((P \,\&\, Q) \leftrightarrow (P \vee Q))$
14. Consider the following rule:
    RAA (*Reductio ad absurdum*) ϕ may be entered on a line if ψ and $-\psi$ appear on earlier lines; as premise-numbers of the new line take all those of the aforementioned earlier lines, with the exception (if desired) of any that is the line number of a line on which $-\phi$ appears.
    Show that, for any sentence ϕ and set of sentences Γ, ϕ is derivable from Γ by the rules P, MP, RAA, C, D if and only if ϕ is derivable from Γ by the rules P, MP, MT, C, D.
15. Show that if $-\phi$ is a tautologous SC sentence, then ϕ contains at least one occurrence of the negation sign '−'.
16. A sentence ϕ is a *truth-functional component* (TFC) of a sentence ψ if and only if ϕ occurs free in ψ at least once. A set of sentences Γ is a *truth-set* for a sentence ϕ if and only if every element of Γ is a truth-functional component of ϕ and Γ further satisfies the following conditions: for all sentences ψ, χ
    1) if $-\psi$ is a TFC of ϕ, then $-\psi \in \Gamma$ iff $\psi \notin \Gamma$
    2) if $(\psi \vee \chi)$ is a TFC of ϕ, then $(\psi \vee \chi) \in \Gamma$ iff $\psi \in \Gamma$ or $\chi \in \Gamma$
    3) if $(\psi \,\&\, \chi)$ is a TFC of ϕ, then $(\psi \,\&\, \chi) \in \Gamma$ iff $\psi \in \Gamma$ and $\chi \in \Gamma$
    4) if $(\psi \to \chi)$ is a TFC of ϕ, then $(\psi \to \chi) \in \Gamma$ iff $\psi \notin \Gamma$ or $\chi \in \Gamma$
    5) if $(\psi \leftrightarrow \chi)$ is a TFC of ϕ, then $(\psi \leftrightarrow \chi) \in \Gamma$ iff $\psi \in \Gamma$ iff $\chi \in \Gamma$
    Show that, for any sentence ϕ, ϕ is a tautology if and only if ϕ belongs to all of its truth-sets.

7

RULES OF INFERENCE FOR $\mathfrak{L}$

In this chapter we present a system of six rules by means of which it is possible to derive from any given set of sentences Γ those and only those sentences that are consequences of Γ. The rules are first stated as carefully as is compatible with reasonable conciseness; in the second section they are discussed seriatim and illustrated with numerous examples. The third section states, explains, illustrates, and justifies three so-called short-cut rules. Although these are very useful, they are in principle dispensable, since the six basic rules are sufficient by themselves. The fourth section offers some suggestions for constructing derivations of given conclusions from given premises. In the last section we devote our attention to the so-called theorems of logic, i.e., to sentences derivable from the empty set. Various theorems are demonstrated, and others are stated without proof.

1. *Basic rules; derivability.* We have seen that the truth-table method provides a way of deciding whether or not any given sentence is a tautology, and whether or not it is a tautological consequence of a given finite set of other sentences. In many cases the method is cumbersome, but nevertheless it does exist. When we turn from the tautologies to the valid sentences generally, however, we find that matters are quite otherwise. We should like to have a similar procedure that would always allow us (or some machine) to produce a yes-or-no answer to the question whether a given sentence is valid. Obviously the truth-table method no longer suffices. If it

informs us that a given sentence is tautologous, then of course that sentence is established as valid, for all tautologies are valid; but if it informs us that the given sentence is not tautologous we cannot conclude that it is invalid. Thus, although '$(x)Fx \rightarrow Fa$' is not tautologous it is none the less valid, and of course there are infinitely many similar cases.

It turns out that not only has no such method been found up to now, but no such method *can ever* be found. The impossibility of a decision procedure for validity was proved in 1936 by the American logician Alonzo Church. It implies, of course, that there is also no step-by-step procedure for deciding whether or not a given sentence is a consequence of other sentences. The picture is not entirely dark, however, for we do possess something analogous to the system of inference rules given in section 6 of the preceding chapter. It is possible to state a few mechanically applicable rules which together have the property that a sentence ϕ will be a consequence of a set of sentences Γ if and only if there is a correct derivation of ϕ from the sentences Γ, where a 'correct' derivation is one in which each step is made in accord with the stated rules. The absence of a decision procedure makes the existence of such rules much more important than it was in the case of the tautologies.

Again we give for reference an exact statement of the rules, followed by various examples and explanatory comments.

A *derivation* is a finite sequence of consecutively numbered lines, each consisting of a sentence together with a set of numbers (called the *premise-numbers* of the line), the sequence being constructed according to the following rules (in these statements ϕ and ψ are arbitrary formulas, α is a variable, and β is an individual constant):

P (Introduction of premises) Any sentence may be entered on a line, with the line number taken as the only premise-number.

T (Tautological inference) Any sentence may be entered on a line if it is a tautological consequence of a set of sentences that appear on earlier lines;* as premise-numbers of the new line take all premise-numbers of those earlier lines.

C (Conditionalization) The sentence $(\phi \rightarrow \psi)$ may be entered on a line if ψ appears on an earlier line; as premise-numbers of the new line take all those of the earlier line, with the exception (if desired) of any that is the line number of a line on which ϕ appears.

US (Universal specification) The sentence $\phi\ \alpha/\beta$ may be entered on a line if $(\alpha)\phi$ appears on an earlier line; as premise-numbers of the new line take those of the earlier line.

*As in the sentential calculus system of Chapter 6, to say that a sentence ϕ appears on a line is to say that the line consists of ϕ and a set of numbers; to say that a sentence ϕ is a premise of a given line is to say that ϕ appears on a line numbered by a premise-number of the given line.

UG (Universal generalization) The sentence $(\alpha)\phi$ may be entered on a line if $\phi\ \alpha/\beta$ appears on an earlier line and β occurs neither in ϕ nor in any premise of that earlier line;* as premise-numbers of the new line take those of the earlier line.

E (Existential quantification) The sentence $(\exists\alpha)\phi$ may be entered on a line if $-(\alpha)-\phi$ appears on an earlier line, or vice versa; as premise-numbers of the new line take those of the earlier line.

A derivation in which a sentence ϕ appears on the last line and all premises of that line belong to a set of sentences Γ is called a *derivation* (*or proof*) *of* ϕ *from* Γ.

A sentence ϕ is *derivable from* the set of sentences Γ if and only if there is a derivation of ϕ from Γ.

Obviously, since any line of a derivation has only finitely many premises, if a sentence is derivable from an infinite set of sentences Γ it is derivable from a finite subset of Γ.

2. *Examples and explanations.* Here are three examples of derivations:

I. The following is a derivation of the sentence '$(x)(Fx \rightarrow Hx)$' from the set consisting of the two sentences '$(x)(Fx \rightarrow Gx)$' and '$(x)(Gx \rightarrow Hx)$'. Immediately to the right of the derivation are various citations. Strictly speaking these are not part of the derivation; they are included only to make it easier for the reader to check that the rules have been followed. Since in the present development we shall employ no rule similar to rule TH of the previous chapter, we are able without ambiguity to drop parentheses when citing line numbers.

{1}	(1) $(x)(Fx \rightarrow Gx)$	P
{2}	(2) $(x)(Gx \rightarrow Hx)$	P
{3}	(3) Fa	P
{1}	(4) $Fa \rightarrow Ga$	1 US
{1,3}	(5) Ga	3,4 T
{2}	(6) $Ga \rightarrow Ha$	2 US
{1,2,3}	(7) Ha	5,6 T
{1,2}	(8) $Fa \rightarrow Ha$	3,7 C
{1,2}	(9) $(x)(Fx \rightarrow Hx)$	8 UG

Perhaps the following will serve as an intuitive paraphrase of the argument formally presented in the derivation just given. Assume that whatever has the property F has the property G; and assume that whatever has the property G has the property H. Now let a be an arbitrarily chosen ob-

*As in the sentential calculus system of Chapter 6, to say that a sentence ϕ appears on a line is to say that the line consists of ϕ and a set of numbers; to say that a sentence ϕ is a premise of a given line is to say that ϕ appears on a line numbered by a premise-number of the given line.

ject, and suppose that it has the property F. By (1), if a has F, then it has G; so by (3) and (4), a has G. But, by (2), if a has G, then it has H; so by (5) and (6) a has H. This has been derived from (1), (2) and (3). Hence, from (1) and (2) alone it follows that if a has F then a has H. But a was an arbitrarily chosen object; hence, *any* object that has F has H.

II. A shorter derivation of the same conclusion from the same premises runs as follows:

{1}	(1)	$(x)(Fx \rightarrow Gx)$	P
{2}	(2)	$(x)(Gx \rightarrow Hx)$	P
{1}	(3)	$Fa \rightarrow Ga$	1 US
{2}	(4)	$Ga \rightarrow Ha$	2 US
{1,2}	(5)	$Fa \rightarrow Ha$	3,4 T
{1,2}	(6)	$(x)(Fx \rightarrow Hx)$	5 UG

Note that the premises of each line are the sentences appearing on the lines numbered by the premise-numbers of that line, so that the premises of line (6) are the sentences appearing on lines (1) and (2), i.e., the sentences '$(x)(Fx \rightarrow Gx)$' and '$(x)(Gx \rightarrow Hx)$'.

III. As a third example consider a derivation of '$(\exists x)(Hx \& -Fx)$' from the set consisting of '$(x)(Fx \rightarrow Gx)$' and '$(\exists x)(Hx \& -Gx)$'.

{1}	(1)	$(x)(Fx \rightarrow Gx)$	P
{2}	(2)	$(\exists x)(Hx \& -Gx)$	P
{3}	(3)	$(x)-(Hx \& -Fx)$	P
{3}	(4)	$-(Ha \& -Fa)$	3 US
{1}	(5)	$Fa \rightarrow Ga$	1 US
{1,3}	(6)	$-(Ha \& -Ga)$	4,5 T
{1,3}	(7)	$(x)-(Hx \& -Gx)$	6 UG
{1}	(8)	$(x)-(Hx \& -Fx) \rightarrow (x)-(Hx \& -Gx)$	3,7 C
{1}	(9)	$-(x)-(Hx \& -Gx) \rightarrow -(x)-(Hx \& -Fx)$	8 T
{2}	(10)	$-(x)-(Hx \& -Gx)$	2 E
{1,2}	(11)	$-(x)-(Hx \& -Fx)$	9,10 T
{1,2}	(12)	$(\exists x)(Hx \& -Fx)$	11 E

Before further examples are presented, some explanations of the rules are in order.

Rule P. The purpose of this rule is to allow us to introduce premises into a derivation. While ordinarily premises occur at the beginning of a derivation, this is not necessary, for a derivation may begin with an application of rule T (see below) and premises may be brought in wherever and whenever convenient. Sometimes, as in examples I and III above, it is advantageous to assume certain premises in addition to those given; after these have served their purpose they may be discharged by means of rule

C. Note that any line justified by rule P will itself have one premise, namely, the sentence appearing on the line. Note also that only *sentences* may be entered according to this rule. Indeed, none of the rules allows us to enter a formula that is not a sentence.

Rule T. If ϕ is a sentence that is a tautological consequence of a set of sentences Γ, rule T allows us to enter ϕ on a new line provided that all members of Γ appear on earlier lines. Thus by this rule we can write down tautological consequences of what we already have. It is not excluded that Γ may be the empty set; in that case ϕ must be a tautology. Hence a derivation can begin with a tautology, as in the following one-line derivation:

Λ (1) $Fa \vee -Fa$ T

for the sentence '$Fa \vee -Fa$' is a tautological consequence of the empty set of sentences and each member of the empty set appears on an earlier line of this derivation. In applications of rule T the set Γ is always finite, since, of course, any given line of a derivation is preceded by only a finite number of lines. Therefore, one can check the correctness of a purported application of rule T by constructing the corresponding conditional and testing it for tautologousness. For instance, in connection with line (6) of example III the corresponding conditional is

$$(-(Ha \,\&\, -Fa) \,\&\, (Fa \rightarrow Ga)) \rightarrow -(Ha \,\&\, -Ga),$$

which is tautologous. It is obvious that a large repertoire of tautologies is indispensable for virtuosity in the application of this rule.

Rule C. In its principal use, rule C says that if you have succeeded in obtaining ψ from premises that include a sentence ϕ, then you may infer $\phi \rightarrow \psi$ and drop ϕ from your list of premises. In other words, the price of getting rid of the premise ϕ is that it must be added as an antecedent to ψ. Thus the rule is helpful in deriving conditionals: assume the antecedent of the desired conditional, derive the consequent, and then conditionalize to obtain the conditional (see, for instance, lines (3)–(8) of example I). For the sake of brevity the rule has been formulated in a manner permitting one also to add an antecedent ϕ that is not a premise of the given line; this inference is already permitted by rule T, however, since $\phi \rightarrow \psi$ is a tautological consequence of ψ. The clause 'if desired' is included in order not to depart unnecessarily from the language of the corresponding SC rule, where such a proviso was requisite to give the system the property mentioned after the proof of SC theorem 10 (see page 100).

Experience shows that the most usual mistake in the application of this rule is that of dropping the *premise* numbers, and not the *line* number, of the line being discharged.

Rule US. This rule takes us from a universal generalization to a particular case. Students encounter difficulty in its application only if they are unable to recognize a universal generalization or they do not understand how to form particular cases of such generalizations. Thus the rule is *not* applicable to the sentence

$$(x)(Fx \rightarrow Ga) \rightarrow Ga,$$

nor to the sentence

$$-(x)Fx,$$

since neither of these is the universal generalization of any formula. It will, however, take us from

$$(x)Fax$$

to

$$Fab$$

and also to

$$Faa.$$

Note a limiting case: the rule allows us to drop a vacuous universal quantifier. For example, it allows transition from

$$(x)Fa$$

to

$$Fa,$$

where we consider '*Fa*' as the result of replacing all free occurrences of '*x*' in '*Fa*' by occurrences of '*b*'. Of course, in any application of this rule, or of the other rules, too, one must be clear about what values one is taking for the variables 'α', 'β', and 'ϕ'.

Thus in line (4) of example I, US is applied by taking $\alpha =$ '*x*', $\phi =$ '$(Fx \rightarrow Gx)$', $\beta =$ '*a*'; thus $(\alpha)\phi =$ '$(x)(Fx \rightarrow Gx)$', and $\phi\ \alpha/\beta =$ '$(Fa \rightarrow Ga)$'. The application of US in line (4) of example III is accomplished by taking $\alpha =$ '*x*', $\phi =$ '$-(Hx\ \&\ -Fx)$', $\beta =$ '*a*', so that $(\alpha)\phi =$ '$(x) - (Hx\ \&\ -Fx)$' and $\phi\ \alpha/\beta =$ '$-(Ha\ \&\ -Fa)$'.

Rule UG. Just as the rule US enables us to pass from a universal generalization to a particular case, the rule UG takes us from the particular case back to the universal generalization. But this time some restrictions are needed. Although any particular case $\phi\ \alpha/\beta$ is a consequence of the corresponding generalization $(\alpha)\phi$, the converse does not always hold; hence we cannot expect to have a rule UG which simply reverses rule US. Instead, we construct a rule on the basis of the following fact: if $\phi\ \alpha/\beta$ is a consequence of Γ and the individual constant β occurs neither in ϕ nor in any member of Γ, then $(\alpha)\phi$ is a consequence of Γ. Very loosely, this says that whatever can be established about a given individual (named by β) on the basis of assumptions (Γ) not specifically mentioning that individual, can be established (on the basis of the same assumptions) about every individual. For instance, '$Fa \rightarrow Ha$' follows from '$(x)(Fx \rightarrow Gx)$' and '$(x)(Gx \rightarrow Hx)$', but then so does '$(x)(Fx \rightarrow Hx)$' (see examples I and II).

On the other hand, '$Fa \rightarrow Ha$' follows from '$-Fa$', but '$(x)(Fx \rightarrow Hx)$' does not; here the assumption '$-Fa$' contains the constant 'a'.

On the basis of considerations of this sort we are justified in employing a rule allowing us to generalize any conclusion obtained from premises that do not contain the individual constant to be generalized upon. The further condition, that the constant generalized upon shall no longer occur in the result of the generalization, serves to make sharper the intuitive notion that '$(\alpha)\phi$ says of every object what $\phi\ \alpha/\beta$ says of the object denoted by β'. We have to block such inferences as

{1}	(1) $(x)Fxx$	P
{1}	(2) Faa	1 US
{1}	(3) $(x)Fxa$	2 UG (erroneous),

which are clearly invalid, even though someone might think that '$(x)Fxa$' says of everything what 'Faa' says of the object denoted by 'a' (namely, that it stands in the relation F to a). The rule as formulated does not permit this inference, since the constant β (='a') occurs in ϕ (='Fxa'). Instead of '$(x)Fxa$', however, we could legitimately have obtained '$(x)Fxx$' again, or, as another limiting case, the sentence '$(x)Faa$'.

In understanding the rule it may be helpful to observe the following: the stipulation that β does not occur in ϕ could have been replaced by the equivalent stipulation that $\phi\ \alpha/\beta$ shall have occurrences of β where *and only where* ϕ has free occurrences of α. Thus, while US takes us from a sentence $(\alpha)\phi$ to any sentence ϕ' that has occurrences of a constant β wherever ϕ has free occurrences of α, the rule UG takes us from a sentence ϕ' to $(\alpha)\phi$ if ϕ' has occurrences of β where *and only where* ϕ has free occurrences of α (provided also that β occurs in no premise of the line generalized upon).

Rule E. To say that there is at least one object satisfying a certain condition is equivalent to saying that not every object fails to satisfy the condition. Rule E accordingly allows passage from $(\exists\alpha)\phi$ to $-(\alpha)-\phi$ and vice versa. It may be applied only when the sentence $(\exists\alpha)\phi$, and therefore also the sentence $-(\alpha)-\phi$, is the *entire* sentence appearing on the line, and not just a part of it. Thus

{1}	(1) $(\exists x)(Fx \,\&\, Gx)$	P
{1}	(2) $-(x)-(Fx \,\&\, Gx)$	1 E

is correct, but

{1}	(1) $-(\exists x)(Fx \,\&\, Gx)$	P
{1}	(2) $--(x)-(Fx \,\&\, Gx)$	1 E (erroneous)

is not, nor is

{1}	(1) $(\exists x)Fx \,\&\, Ga$	P
{1}	(2) $-(x)-Fx \,\&\, Ga$	1 E (erroneous)

Now for some further examples:

IV.

{1}	(1) $(x)(y)((Fx \& Gy) \rightarrow -Hxy)$	P
{2}	(2) $(x)(Ix \rightarrow Gx)$	P
{1}	(3) $(y)((Fa \& Gy) \rightarrow -Hay)$	1 US
{1}	(4) $(Fa \& Gb) \rightarrow -Hab$	3 US
{2}	(5) $Ib \rightarrow Gb$	2 US
{1,2}	(6) $(Fa \& Ib) \rightarrow -Hab$	4,5 T
{1,2}	(7) $(y)((Fa \& Iy) \rightarrow -Hay)$	6 UG
{1,2}	(8) $(x)(y)((Fx \& Iy) \rightarrow -Hxy)$	7 UG

In this derivation the strategy was to drop quantifiers from the premises by means of rule US, operate on the results by means of rule T, and then employ rule UG to replace quantifiers as needed. This strategy is applicable in a large number of cases. Note that it took two steps to remove the quantifiers from the sentence on line (1).

V.

{1}	(1) $(x)(y)(Fxy \rightarrow -Fyx)$	P
{1}	(2) $(y)(Fay \rightarrow -Fya)$	1 US
{1}	(3) $Faa \rightarrow -Faa$	2 US
{1}	(4) $-Faa$	3 T
{1}	(5) $(x)-Fxx$	4 UG

VI.

{1}	(1) $(x)(y)(z)((Fxy \& Fyz) \rightarrow Fxz)$	P
{1}	(2) $(y)(z)((Fay \& Fyz) \rightarrow Faz)$	1 US
{1}	(3) $(z)((Fab \& Fbz) \rightarrow Faz)$	2 US
{1}	(4) $(Fab \& Fba) \rightarrow Faa$	3 US
{5}	(5) $(x)(y)(Fxy \rightarrow Fyx)$	P
{5}	(6) $(y)(Fay \rightarrow Fya)$	5 US
{5}	(7) $Fab \rightarrow Fba$	6 US
{8}	(8) $-Faa$	P
{1,5,8}	(9) $-Fab$	4,7,8 T
{1,5,8}	(10) $(y)-Fay$	9 UG
{1,5}	(11) $-Faa \rightarrow (y)-Fay$	8,10 C
{1,5}	(12) $(x)(-Fxx \rightarrow (y) -Fxy)$	11 UG

The preceding two derivations show that one must not overlook the possibility of using US to drop two different quantifiers to the same constant.

VII. In the following derivation, as in VI, there is a relatively complicated application of rule T; if desired, any such application of this rule can be replaced by a series of simpler applications.

{1}	(1) $(x)(Fx \rightarrow (Gx \vee Hx))$	P
{2}	(2) $(x)((Fx \& Gx) \rightarrow Ix)$	P
{3}	(3) $(x)((Fx \& Hx) \rightarrow Jx)$	P
{1}	(4) $Fa \rightarrow (Ga \vee Ha)$	1 US
{2}	(5) $(Fa \& Ga) \rightarrow Ia$	2 US
{3}	(6) $(Fa \& Ha) \rightarrow Ja$	3 US
{1,2,3}	(7) $(Fa \& -Ja) \rightarrow Ia$	4,5,6 T
{1,2,3}	(8) $(x)((Fx \& -Jx) \rightarrow Ix)$	7 UG

VIII. Note that in the next derivation it is necessary to apply rule T to line (1) before the quantifier can be dropped by rule US.

{1}	(1) $Fa \& (x)(Fx \rightarrow Gax)$	P
{1}	(2) $(x)(Fx \rightarrow Gax)$	1 T
{1}	(3) $Fa \rightarrow Gaa$	2 US
{1}	(4) Gaa	1,3 T

IX. The following derivations should be re-examined after the introduction of rule EG in the next section.

{1}	(1) Fa	P
{2}	(2) $(x)-Fx$	P
{2}	(3) $-Fa$	2 US
Λ	(4) $(x)-Fx \rightarrow -Fa$	2,3 C
{1}	(5) $-(x)-Fx$	1,4 T
{1}	(6) $(\exists x)Fx$	5 E

X.

{1}	(1) $(x)(Fx \rightarrow Gx)$	P
{2}	(2) $(x)(Fx \vee Hx)$	P
{3}	(3) $(x)(Hx \rightarrow Fx)$	P
{1}	(4) $Fa \rightarrow Ga$	1 US
{2}	(5) $Fa \vee Ha$	2 US
{3}	(6) $Ha \rightarrow Fa$	3 US
{1,2,3}	(7) $Fa \& Ga$	4,5,6 T
{8}	(8) $(x)-(Fx \& Gx)$	P
{8}	(9) $-(Fa \& Ga)$	8 US
Λ	(10) $(x)-(Fx \& Gx) \rightarrow -(Fa \& Ga)$	8,9 C
{1,2,3}	(11) $-(x)-(Fx \& Gx)$	7,10 T
{1,2,3}	(12) $(\exists x)(Fx \& Gx)$	11 E

In derivation X the strategy has involved assuming (in line (8)) something equivalent to the contradictory* of the desired conclusion and then deriv-

*A sentence and its negation are called *contradictories* of one another. Though any sentence of the form $-\phi$ has two equivalent contradictories, ϕ and $--\phi$, it has become customary to speak loosely of 'the' contradictory of a sentence.

ing (in line (9)) the contradictory of one of the lines already obtained. This technique is sometimes called the method of *reductio ad absurdum:* assume the contradictory of what you wish to show, and then seek to derive the contradictory of a tautology or of one of your premises or of some sentence you have already derived from the premises. If you are successful in doing this, application of C and T (as in lines (10) and (11) of X) will give the desired conclusion.

3. *The short-cut rules EG, ES, and Q.* Another inference rule often employed in deduction systems of this type runs as follows:

EG (Existential generalization) The sentence $(\exists\alpha)\phi$ may be entered on a line if $\phi\ \alpha/\beta$ appears on an earlier line; as premise-numbers of the new line take those of the earlier line.

With the help of this rule the derivations IX and X of the previous section could each have been shortened by four lines. IX would have looked like this:

{1}	(1) *Fa*	P
{1}	(2) $(\exists x)Fx$	1 EG

and X could have ended as follows:

{1,2,3}	(7) *Fa* & *Ga*	4,5,6 T
{1,2,3}	(8) $(\exists x)(Fx\ \&\ Gx)$	7 EG

In view of its convenience we shall add this rule to our list, bearing in mind, however, that it is only a 'short-cut' rule and in principle superfluous. Any inference that can be made with its help can also be made on the basis of the rules P, T, C, US, UG and E alone. For, if $\phi\ \alpha/\beta$ appears on the *i*-th line of a derivation, with premise-numbers $n_1, \ldots, n_p$, we can always get $(\exists\alpha)\phi$ on a later line (and with the same premise-numbers) via the following route:

$\{n_1, \ldots, n_p\}$	(i)	$\phi\ \alpha/\beta$	—
.	.	.	.
.	.	.	.
.	.	.	.
$\{j\}$	(j)	$(\alpha)-\phi$	P
$\{j\}$	$(j+1)$	$-\phi\ \alpha/\beta$	j US
Λ	$(j+2)$	$(\alpha)-\phi \rightarrow -\phi\ \alpha/\beta$	$j, j+1$ C
$\{n_1, \ldots, n_p\}$	$(j+3)$	$-(\alpha)-\phi$	$i, j+2$ T
$\{n_1, \ldots, n_p\}$	$(j+4)$	$(\exists\alpha)\phi$	$j+3$ E

The student should notice that the sequence of steps here described in general is just the sequence that constitutes the last six lines of derivations IX and X. For, if we take $\phi =$ '*Fx*', $\alpha =$ '*x*', $\beta =$ '*a*', $i = 1$, $j = 2$, and

$n_1 = 1$, we get lines (1)–(6) of IX, while by taking ϕ = '*Fx* & *Gx*', α = '*x*', β = '*a*', $i = 7$, $j = 8$, $n_1 = 1$, $n_2 = 2$, $n_3 = 3$, we get lines (7)–(12) of X. In a similar way *any* application of EG can be eliminated at the price of four extra lines.

The rule EG is in a manner 'dual' to US. It will take us from

$$Faa$$

to

$$(\exists x)Fxx$$

or to

$$(\exists x)Fax$$

or to

$$(\exists x)Fxa.$$

Also, as a limiting case, it will allow the move from

$$Faa$$

to

$$(\exists x)Faa,$$

where ϕ = '*Faa*', α = '*x*', and β = '*b*' or any other individual constant.

An even more useful short-cut rule is the rule ES. To prepare himself for understanding the application of this rule the student should first compare the following two derivations of the sentence '$(\exists x)Gx$' from the sentences '$(x)(Fx \rightarrow Gx)$' and '$(\exists x)Fx$'.

XI.

{1}	(1) $(x)(Fx \rightarrow Gx)$	P
{2}	(2) $(\exists x)Fx$	P
{2}	(3) $-(x)-Fx$	2 E
{4}	(4) $(x)-Gx$	P
{1}	(5) $Fa \rightarrow Ga$	1 US
{4}	(6) $-Ga$	4 US
{1,4}	(7) $-Fa$	5,6 T
{1,4}	(8) $(x)-Fx$	7 UG
{1}	(9) $(x)-Gx \rightarrow (x)-Fx$	4,8 C
{1,2}	(10) $-(x)-Gx$	3,9 T
{1,2}	(11) $(\exists x)Gx$	10 E

XII.

{1}	(1) $(x)(Fx \rightarrow Gx)$	P
{2}	(2) $(\exists x)Fx$	P
{3}	(3) Fa	P
{1}	(4) $Fa \rightarrow Ga$	1 US
{1,3}	(5) Ga	3,4 T
{1,3}	(6) $(\exists x)Gx$	5 EG
{1,2}	(7) $(\exists x)Gx$	2,3,6 ES

Derivation XII may be very roughly paraphrased as follows: Assume that whatever has F has G, and assume that something has F. Call that something '*a*', i.e., assume that *a* has F. Now, by (1), if *a* has F then *a* has G; so, by (3) and (4), *a* has G. Therefore something has G. But this conclusion has been derived from (1) and (3), and in (3) our choice of '*a*' was inessentian—we could have used '*b*' or any other name; hence the conclusion follows equally well from (1) and (2).

The crucial steps are (2), (3), and (6). Given, in line (2), that something has the property F, we decide to give it a name for the duration of the argument; we choose '*a*', since that constant has not been used up to this point. Now, it would not do to *derive* 'Fa' from '$(\exists x)Fx$', since 'Fa' is certainly *not* a consequence of '$(\exists x)Fx$'. Instead, we take 'Fa' as a new premise; we are saying, in effect, '*let a* be one of the objects that has F'. In the subsequent argument, however, we make no special use of the fact that we chose '*a*' instead of some other constant, and eventually we reach a conclusion in which '*a*' is no longer involved. Under such circumstances rule ES allows us to transfer the dependence of this conclusion from 'Fa' to the weaker assertion '$(\exists x)Fx$'.

Thus a typical application of the rule takes place in a situation in which we have an existential sentence occurring on the *i*-th line of a derivation:

$\{n_1, \ldots, n_p\}$ (i) $(\exists\alpha)\psi$ —

and later add a premise:

$\{j\}$ (j) $\psi\ \alpha/\beta$ P,

which amounts to saying 'call it β', and still later reach a conclusion:

$\{m_1, \ldots, m_q, j\}$ (k) ϕ —.

Under certain provisos the rule then allows us to replace dependence on line (j) by dependence on (the premises of) line (i); i.e., to add a line:

$\{m_1, \ldots, m_q, n_1, \ldots, n_p\}$ (l) ϕ i, j, k ES.

The reasons for the provisos will become evident when we show how an application of the short-cut rule ES can always in principle be avoided (at the cost, of course, of a number of additional steps).

Here, then, is the rule:

ES (Existential specification) Suppose that $(\exists\alpha)\psi$ appears on line i of a derivation, that $\psi\ \alpha/\beta$ appears (as a premise) on a later line j, and that ϕ appears on a still later line k; and suppose further that the constant β occurs neither in ϕ, ψ, nor in any premise of line k other than $\psi\ \alpha/\beta$; then ϕ may be entered on a new line. As premise-numbers of the new line take all those of lines i and k, except the number j.

To justify this rule, which should be compared with statement 13) on page 66 above, we observe that if its various suppositions are satisfied we may proceed as follows to obtain the inference it permits:

$\{n_1, \ldots, n_p\}$	(i)	$(\exists\alpha)\ \psi$	—
.	.	.	.
.	.	.	.
$\{j\}$	(j)	$\psi\ \alpha/\beta$	P
.	.	.	.
.	.	.	.
$\{m_1, \ldots, m_q, j\}$	(k)	ϕ	—
$\{m_1, \ldots, m_q\}$	$(k+1)$	$\psi\ \alpha/\beta \rightarrow \phi$	j,k C
$\{k+2\}$	$(k+2)$	$-\phi$	P
$\{m_1, \ldots, m_q, k+2\}$	$(k+3)$	$-\psi\ \alpha/\beta$	$k+1,k+2$ T
$\{m_1, \ldots, m_q, k+2\}$	$(k+4)$	$(\alpha)-\psi$	$k+3$ UG
$\{m_1, \ldots, m_q\}$	$(k+5)$	$-\phi \rightarrow (\alpha)-\psi$	$k+2,k+4$ C
$\{n_1, \ldots, n_p\}$	$(k+6)$	$-(\alpha)-\psi$	i E
$\{m_1, \ldots, m_q, n_1, \ldots, n_p\}$	$(k+7)$	ϕ	$k+5,k+6$ T

Note line $(k+4)$. It is here that the various provisos about β are used. For if UG is to be correctly applied, β must occur neither in $(\alpha)-\psi$ nor in any premise of line $(k+3)$; this amounts to the requirement that β not occur in ϕ, ψ, or in any premise of line (k) other than $\psi\ \alpha/\beta$.

It should be mentioned that if the number j is not among the premise-numbers of line (k) (and the rule does not suppose that it is), then line $(k+1)$ will follow from line k alone, by rule C.

The standard use of rule ES is part of what we may call the 'ES strategy': if in constructing a derivation you have reached a sentence $(\exists\alpha)\ \psi$, try adding as a new premise $\psi\ \alpha/\beta$, choosing for β an individual constant not yet used in the derivation and not occurring in the desired conclusion ϕ. Next try to derive ϕ without adding further premises involving β. If you are successful, rule ES will allow you to discharge the premise $\psi\ \alpha/\beta$ in favor of the premises of $(\exists\alpha)\psi$.

Assuming for the moment that the basic rules P through E are sound (will never give us a false conclusion from true premises), the above description of how any ES inference can be made by using only rules P–E shows that ES is sound, too. Thus the provisos about β are sufficient to block undesirable inferences. That they are also necessary may be seen from the following three 'derivations' in which ES is erroneously applied. In each of them the sentence appearing on the last line is not a consequence of its premises, yet would be derivable if the appropriate restriction were omitted from the statement of ES.

{1}	(1) $(\exists x)Fxa$	P	(i) $(\exists\alpha)\ \psi$
{2}	(2) Faa	P	(j) $\psi\ \alpha/\beta$
{2}	(3) $(\exists x)Fxx$	2 EG	(k) ϕ

{1}	(4) $(\exists x)Fxx$	1,2,3 ES	(erroneous, because 'a' ($=\beta$) occurs in 'Fxa' ($=\psi$))

{1}	(1) $(\exists x)Fxx$	P	(i) $(\exists\alpha)\psi$
{2}	(2) Faa	P	(j) $\psi\,\alpha/\beta$
{2}	(3) $(\exists x)Fxa$	2 EG	(k) ϕ
{1}	(4) $(\exists x)Fxa$	1,2,3 ES	(erroneous, because 'a' ($=\beta$) occurs in '$(\exists x)Fxa$' ($=\phi$))

{1}	(1) $(\exists x)Fx$	P	(i) $(\exists\alpha)\psi$
{2}	(2) Fa	P	(j) $\psi\,\alpha/\beta$
{3}	(3) $-Fa$	P	
{2,3}	(4) $Fa\ \&\ -Fa$	2,3 T	
{2,3}	(5) $(\exists x)(Fx\ \&\ -Fx)$	4 EG	(k) ϕ
{1,3}	(6) $(\exists x)(Fx\ \&\ -Fx)$	1,2,5 ES	(erroneous, because 'a' ($=\beta$) occurs in '$-Fa$', which is a premise of line (5))

We complete our discussion of rule ES by giving further examples of its use.

XIII.

{1}	(1) $(x)(Fx \to (y)(Gy \to Hxy))$	P
{2}	(2) $(x)(Fx \to (z)(Iz \to -Hxz))$	P
{3}	(3) $(\exists x)Fx$	P
{4}	(4) Fa	P
{1}	(5) $Fa \to (y)(Gy \to Hay)$	1 US
{2}	(6) $Fa \to (z)(Iz \to -Haz)$	2 US
{1,4}	(7) $(y)(Gy \to Hay)$	4,5 T
{2,4}	(8) $(z)(Iz \to -Haz)$	4,6 T
{1,4}	(9) $Gb \to Hab$	7 US
{2,4}	(10) $Ib \to -Hab$	8 US
{1,2,4}	(11) $Gb \to -Ib$	9,10 T
{1,2,4}	(12) $(y)(Gy \to -Iy)$	11 UG
{1,2,3}	(13) $(y)(Gy \to -Iy)$	3,4,12 ES

If in this case we eliminated the use of ES by following the general procedure described above, we would obtain

{1,2}	(13) $Fa \to (y)(Gy \to -Iy)$	4,12 C
{14}	(14) $-(y)(Gy \to -Iy)$	P
{1,2,14}	(15) $-Fa$	13,14 T
{1,2,14}	(16) $(x)-Fx$	15 UG
{1,2}	(17) $-(y)(Gy \to -Iy) \to (x)-Fx$	14,16 C

{3}	(18) $-(x)-Fx$	3 E
{1,2,3}	(19) $(y)(Gy \rightarrow -Iy)$	17,18 T

as the last seven lines of the derivation.

XIV.

{1}	(1) $(\exists x)(\exists y)Fxy$	P
{2}	(2) $(\exists y)Fay$	P
{3}	(3) Fab	P
{3}	(4) $(\exists x)Fxb$	3 EG
{3}	(5) $(\exists y)(\exists x)Fxy$	4 EG
{2}	(6) $(\exists y)(\exists x)Fxy$	2,3,5 ES
{1}	(7) $(\exists y)(\exists x)Fxy$	1,2,6 ES

This example illustrates the application of ES within another application of ES.

XV.

{1}	(1) $(\exists y)(x)Fxy$	P
{2}	(2) $(x)Fxa$	P
{2}	(3) Fba	2 US
{2}	(4) $(\exists y)Fby$	3 EG
{2}	(5) $(x)(\exists y)Fxy$	4 UG
{1}	(6) $(x)(\exists y)Fxy$	1,2,5 ES

The foregoing is a derivation of '$(x)(\exists y)Fxy$' from '$(\exists y)(x)Fxy$'; if, however, we try to construct an analogous derivation of the latter from the former we find ourselves blocked (fortunately, for otherwise our system of rules would be unsound):

{1}	(1) $(x)(\exists y)Fxy$	P
{1}	(2) $(\exists y)Fay$	1 US
{3}	(3) Fab	P
{3}	(4) $(x)Fxb$	3 UG (erroneous)
{3}	(5) $(\exists y)(x)Fxy$	4 EG
{1}	(6) $(\exists y)(x)Fxy$	2,3,5 ES

But compare the following:

XVI.

{1}	(1) $(x)(\exists y)(Fx \& Gy)$	P
{1}	(2) $(\exists y)(Fa \& Gy)$	1 US
{3}	(3) $Fa \& Gb$	P
{1}	(4) $(\exists y)(Fc \& Gy)$	1 US
{5}	(5) $Fc \& Gd$	P
{3,5}	(6) $Fc \& Gb$	3,5 T
{1,3}	(7) $Fc \& Gb$	4,5,6 ES

{1,3}	(8) $(x)(Fx \& Gb)$	7 UG
{1,3}	(9) $(\exists y)(x)(Fx \& Gy)$	8 EG
{1}	(10) $(\exists y)(x)(Fx \& Gy)$	2,3,9 ES

Here is another short-cut rule, the utility of which will be self-evident.

Q (Quantifier exchange) The sentence $-(\exists\alpha)\phi$ may be entered on a line if $(\alpha)-\phi$ appears on an earlier line, or vice versa; similarly for the pairs $\{(\exists\alpha)-\phi, -(\alpha)\phi\}$, $\{-(\exists\alpha)-\phi, (\alpha)\phi\}$, $\{(\exists\alpha)\phi, -(\alpha)-\phi\}$; as premise-numbers of the new line take those of the earlier line.

As with EG and ES, we may justify Q by showing how any inference that can be made with its help can also be made without it. In each of the eight cases the argument proceeds by *reductio ad absurdum* and offers no special difficulty.

Finally, an illustration of the use of the new rule Q:

XVII.

{1}	(1) $-(x)(\exists y)(z)Fxyz$	P
{1}	(2) $(\exists x)-(\exists y)(z)Fxyz$	1 Q
{3}	(3) $-(\exists y)(z)Fayz$	P
{3}	(4) $(y)-(z)Fayz$	3 Q
{3}	(5) $-(z)Fabz$	4 US
{3}	(6) $(\exists z)-Fabz$	5 Q
{3}	(7) $(y)(\exists z)-Fayz$	6 UG
{3}	(8) $(\exists x)(y)(\exists z)-Fxyz$	7 EG
{1}	(9) $(\exists x)(y)(\exists z)-Fxyz$	2,3,8 ES

4. *Strategy*. Although no decision procedure is available and in practice a certain amount of insight is often required for deriving a given conclusion from given premises, one can make some general suggestions that will sometimes help. For instance:

(1) If the desired conclusion is of the form $\phi \rightarrow \psi$, add ϕ to the given premises, deduce ψ, and apply C.

(2) If one of the premises is of the form $\phi \vee \psi$, and the desired conclusion is χ, use the preceding method to derive the conditionals $\phi \rightarrow \chi$ and $\psi \rightarrow \chi$, and then apply T to obtain χ.

(3) If the desired conclusion is of the form $\phi \leftrightarrow \psi$, derive $\phi \rightarrow \psi$ and $\psi \rightarrow \phi$, and apply T.

(4) If the conclusion is of the form $(\alpha)\phi$, derive $\phi\ \alpha/\beta$ for some constant β not occurring in the given premises or in ϕ, and apply UG.

(5) If the conclusion is of the form $(\exists\alpha)\phi$, derive $\phi\ \alpha/\beta$ for some constant β and apply EG, or use the technique of *reductio ad absurdum* (i.e., assume $(\alpha)-\phi$ and try to derive the contradictory of one of the given premises or of some sentence derived from the given premises).

(6) Drop universal and existential quantifiers from the premises by using US and the ES strategy; operate on the resulting sentences by rule T; and put the quantifiers back by rules UG and EG. If a premise is molecular with general parts, use the rules T or P to separate the parts before attempting to apply US and the ES strategy.

(7) Whenever the premises seem inadequate, try taking as an additional premise the contradictory of the desired conclusion, with a view toward applying the technique of *reductio ad absurdum*.

These suggestions apply to the premises and conclusions of subarguments, of course, as well as to those of the whole argument.

The best advice of all, however, is simply that the student practice using the rules so that he will become thoroughly familiar with what each of them permits and does not permit. This is what is needed in order to 'see' the way from given premises to a given conclusion.

5. *Theorems of logic.* A sentence ϕ is a *theorem of logic* (or *theorem*, for short) if and only if ϕ is derivable from the empty set of sentences.

From this definition, together with the fact (which we have only stated, not proved) that a sentence ϕ is derivable from a set of sentences Γ if and only if ϕ is a consequence of Γ, it follows that ϕ is a theorem of logic if and only if ϕ is valid.

In view of rule T, it is obvious that every tautologous sentence ϕ is a theorem of logic, for it will have a one-line derivation:

Λ (1) ϕ T

But, of course, not every theorem is a tautology. Among the non-tautologous cases are the following (where no derivation is given, the student should attempt to construct one for himself):

1. $(x)(y)Fxy \leftrightarrow (y)(x)Fxy$
2. $(\exists x)(\exists y)Fxy \leftrightarrow (\exists y)(\exists x)Fxy$ (Cp. example XIV)
3. $(\exists x)(y)Fxy \rightarrow (y)(\exists x)Fxy$ (Cp. example XV)
4. $(x)(Fx \,\&\, Gx) \leftrightarrow ((x)Fx \,\&\, (x)Gx)$

{1}	(1)	$(x)(Fx \,\&\, Gx)$
{1}	(2)	$Fa \,\&\, Ga$
{1}	(3)	Fa
{1}	(4)	Ga
{1}	(5)	$(x)Fx$
{1}	(6)	$(x)Gx$
{1}	(7)	$(x)Fx \,\&\, (x)Gx$
Λ	(8)	$(x)(Fx \,\&\, Gx) \rightarrow ((x)Fx \,\&\, (x)Gx)$
{9}	(9)	$(x)Fx \,\&\, (x)Gx$
{9}	(10)	$(x)Fx$
{9}	(11)	$(x)Gx$

{9}	(12)	Fa
{9}	(13)	Ga
{9}	(14)	$Fa \,\&\, Ga$
{9}	(15)	$(x)(Fx \,\&\, Gx)$
Λ	(16)	$((x)Fx \,\&\, (x)Gx) \rightarrow (x)(Fx \,\&\, Gx)$
Λ	(17)	$(x)(Fx \,\&\, Gx) \leftrightarrow ((x)Fx \,\&\, (x)Gx)$

5. $((x)Fx \vee (x)Gx) \rightarrow (x)(Fx \vee Gx)$
6. $(x)(Fx \rightarrow Gx) \rightarrow ((x)Fx \rightarrow (x)Gx)$
7. $(\exists x)(Fx \,\&\, Gx) \rightarrow ((\exists x)Fx \,\&\, (\exists x)Gx)$
8. $(\exists x)(Fx \vee Gx) \leftrightarrow ((\exists x)Fx \vee (\exists x)Gx)$

{1}	(1)	$(\exists x)(Fx \vee Gx)$
{2}	(2)	$-(\exists x)Fx$
{3}	(3)	$Fa \vee Ga$
{2}	(4)	$(x)-Fx$
{2}	(5)	$-Fa$
{2,3}	(6)	Ga
{2,3}	(7)	$(\exists x)Gx$
{1,2}	(8)	$(\exists x)Gx$
{1}	(9)	$-(\exists x)Fx \rightarrow (\exists x)Gx$
{1}	(10)	$(\exists x)Fx \vee (\exists x)Gx$
Λ	(11)	$(\exists x)(Fx \vee Gx) \rightarrow ((\exists x)Fx \vee (\exists x)Gx)$
{12}	(12)	$(\exists x)Fx$
{13}	(13)	Fa
{13}	(14)	$Fa \vee Ga$
{13}	(15)	$(\exists x)(Fx \vee Gx)$
{12}	(16)	$(\exists x)(Fx \vee Gx)$
Λ	(17)	$(\exists x)Fx \rightarrow (\exists x)(Fx \vee Gx)$
{18}	(18)	$(\exists x)Gx$
.	.	.
Λ	(23)	$(\exists x)Gx \rightarrow (\exists x)(Fx \vee Gx)$
Λ	(24)	$(\exists x)(Fx \vee Gx) \leftrightarrow ((\exists x)Fx \vee (\exists x)Gx)$

9. $(x)(P \,\&\, Fx) \leftrightarrow (P \,\&\, (x)Fx)$
10. $(x)(P \vee Fx) \leftrightarrow (P \vee (x)Fx)$
11. $(x)(P \rightarrow Fx) \leftrightarrow (P \rightarrow (x)Fx)$
12. $(\exists x)(P \,\&\, Fx) \leftrightarrow (P \,\&\, (\exists x)Fx)$
13. $(\exists x)(P \vee Fx) \leftrightarrow (P \vee (\exists x)Fx)$
14. $(\exists x)(P \rightarrow Fx) \leftrightarrow (P \rightarrow (\exists x)Fx)$
15. $(x)(Fx \rightarrow Gx) \rightarrow ((\exists x)Fx \rightarrow (\exists x)Gx)$ (Cp. example XII)
16. $(x)(Fx \rightarrow P) \leftrightarrow ((\exists x)Fx \rightarrow P)$

{1}	(1)	$(x)(Fx \rightarrow P)$
{2}	(2)	$(\exists x)Fx$
{3}	(3)	Fa
{1}	(4)	$Fa \rightarrow P$

{1,3}	(5)	P
{1,2}	(6)	P
{1}	(7)	$(\exists x)Fx \to P$
Λ	(8)	$(x)(Fx \to P) \to ((\exists x)Fx \to P)$
{9}	(9)	$-(x)(Fx \to P)$
{9}	(10)	$(\exists x)-(Fx \to P)$
{11}	(11)	$-(Fa \to P)$
{11}	(12)	Fa
{11}	(13)	$(\exists x)Fx$
{11}	(14)	$(\exists x)Fx \,\&\, -P$
{9}	(15)	$(\exists x)Fx \,\&\, -P$
Λ	(16)	$-(x)(Fx \to P) \to ((\exists x)Fx \,\&\, -P)$
Λ	(17)	$(x)(Fx \to P) \leftrightarrow ((\exists x)Fx \to P)$

17. $(\exists x)(Fx \to P) \leftrightarrow ((x)Fx \to P)$
18. $(x)(\exists y)(Fx \,\&\, Gy) \leftrightarrow ((x)Fx \,\&\, (\exists y)Gy)$
19. $(x)(\exists y)(Fx \,\&\, Gy) \leftrightarrow (\exists y)(x)(Fx \,\&\, Gy)$ (Cp. examples XV and XVI)
20. $(x)(\exists y)(Fx \vee Gy) \leftrightarrow ((x)Fx \vee (\exists y)Gy)$

{1}	(1)	$(x)(\exists y)(Fx \vee Gy)$
{2}	(2)	$-(\exists y)Gy$
{2}	(3)	$(y)-Gy$
{1}	(4)	$(\exists y)(Fa \vee Gy)$
{5}	(5)	$Fa \vee Gb$
{2}	(6)	$-Gb$
{2,5}	(7)	Fa
{1,2}	(8)	Fa
{1,2}	(9)	$(x)Fx$
{1}	(10)	$-(\exists y)Gy \to (x)Fx$
Λ	(11)	$(x)(\exists y)(Fx \vee Gy) \to (-(\exists y)Gy \to (x)Fx)$
{12}	(12)	$(x)Fx$
{12}	(13)	Fa
{12}	(14)	$Fa \vee Gb$
{12}	(15)	$(\exists y)(Fa \vee Gy)$
{12}	(16)	$(x)(\exists y)(Fx \vee Gy)$
Λ	(17)	$(x)Fx \to (x)(\exists y)(Fx \vee Gy)$
{18}	(18)	$(\exists y)Gy$
{19}	(19)	Gc
{19}	(20)	$Fa \vee Gc$
{19}	(21)	$(\exists y)(Fa \vee Gy)$
{19}	(22)	$(x)(\exists y)(Fx \vee Gy)$
{18}	(23)	$(x)(\exists y)(Fx \vee Gy)$
Λ	(24)	$(\exists y)Gy \to (x)(\exists y)(Fx \vee Gy)$
Λ	(25)	$(x)(\exists y)(Fx \vee Gy) \leftrightarrow ((x)Fx \vee (\exists y)Gy)$

21. $(x)(\exists y)(Fx \vee Gy) \leftrightarrow (\exists y)(x)(Fx \vee Gy)$

22. $(x)(\exists y)(Fx \rightarrow Gy) \leftrightarrow (\exists y)(x)(Fx \rightarrow Gy)$
23. $((\exists x)Fx \rightarrow (\exists x)Gx) \leftrightarrow (\exists y)(x)(Fx \rightarrow Gy)$
24. $((x)Fx \rightarrow (x)Gx) \leftrightarrow (\exists x)(y)(Fx \rightarrow Gy)$
25. $(\exists x)(\exists y)(Fx \,\&\, -Fy) \leftrightarrow ((\exists x)Fx \,\&\, (\exists x)-Fx)$
26. $(x)(y)(Fx \rightarrow Fy) \leftrightarrow (-(\exists x)Fx \vee (x)Fx)$
27. $((x)Fx \leftrightarrow (\exists x)Gx) \leftrightarrow (\exists x)(\exists y)(z)(w)((Fx \rightarrow Gy) \,\&\, (Gz \rightarrow Fw))$
28. $((\exists x)Fx \rightarrow ((\exists x)Gx \rightarrow (x)Hx)) \leftrightarrow (x)(y)(z)((Fx \,\&\, Gy) \rightarrow Hz)$
29. $(\exists y)(x)(Fy \vee (Fx \rightarrow P))$
30. $-(\exists y)(x)(Fxy \leftrightarrow -Fxx)$

EXERCISES

1. In each group derive the last sentence from the others. You may use any of the rules P, T, C, US, UG, E, EG, ES, or Q.

(a) $(x)(y)Fxy$
$(y)(x)Fxy$

(b) $(x)(y)Fxy$
$(x)(y)Fyx$

(c) $(x)Fx$
$(y)Fy$

(d) $(\exists x)Fx$
$(\exists y)Fy$

(e) $(x)(Fx \,\&\, Gx)$
$(x)Fx \,\&\, (x)Gx$

(f) $(x)Fx \vee (x)Gx$
$(x)(Fx \vee Gx)$

(g) $(x)(Fx \rightarrow (Gx \vee Hx))$
$Ga \leftrightarrow (Ha \,\&\, -Ga)$
$-Fa$

(h) $(x)(y)(z)((Fxy \,\&\, Fyz) \rightarrow Fxz)$
$(x)-Fxx$
$(x)(y)(Fxy \rightarrow -Fyx)$

(i) $(x)(Fx \rightarrow (Gx \vee Hx))$
$-Ga \,\&\, Ia$
$(\exists x)(Fx \,\&\, Ix)$
$Fa \rightarrow Ha$

(j) $(x)(Fx \vee Gx)$
$(\exists x)-Gx$
$(x)(Hx \rightarrow -Fx)$
$(\exists x)-Hx$

(k) $(x)(P \,\&\, Fx)$
$(x)Fx$

(l) $(\exists x)(P \vee Fx)$
$P \vee (\exists x)Fx$

(m) $(x)(Fx \leftrightarrow P)$
$(x)Fa \leftrightarrow Fa$

(n) $(x)(Fx \rightarrow (Gx \vee Hx))$
$-(\exists x)(Fx \,\&\, Hx)$
$(x)(Fx \rightarrow Gx)$

(o) $(\exists x)(\exists y)(Fxy \vee Fyx)$
$(\exists x)(\exists y)Fxy$

(p) $(\exists x)Fx \rightarrow (x)Fx$
$(\exists x)(y)(Fx \leftrightarrow Fy)$

2. Using any of the rules P, T, C, US, UG, E, EG, ES, or Q, derive the following sentence from the empty set.
$((\exists x)(Fx \,\&\, Gx) \vee (\exists x)(Fx \,\&\, -Gx)) \vee ((\exists x)(-Fx \,\&\, Gx) \vee (\exists x)(-Fx \,\&\, -Gx))$

3. (a) Using only rules P, T, C, US, UG, and E, derive '$(\exists x)(Gx \,\&\, Hx)$' from '$(x)(Fx \rightarrow Gx)$' and '$(\exists x)(Fx \,\&\, Hx)$'.
   (b) Using rules ES and EG as well, construct a shorter derivation of the same conclusion from the same premises.

4. (a) Construct a 'derivation' in which the only error is that in an application of UG the constant β occurs in ϕ, and in which the last line is not a consequence of its premises.

(b) Do the same, except that the only error is that in an application of UG the constant β occurs in some premise of the line generalized upon.

5. For each of the five arguments given in exercise 3, page 86, derive the symbolized conclusion from the symbolized premises.
6. Give the eight arguments that were said on page 126 'to offer no special difficulty'.
7. For each of the following, either give an interpretation to show that it is not valid, or establish its validity by deriving it from the empty set.
   (a) $((x)Fx \,\&\, (\exists y)Gy) \rightarrow (\exists y)(Fy \,\&\, Gy)$
   (b) $(x)(Fx \vee Gx) \rightarrow ((x)Fx \vee (x)Gx)$
   (c) $(x)(y)(Fxy \rightarrow -Fyx) \rightarrow (x) -Fxx$
   (d) $((\exists x)Fx \leftrightarrow P) \rightarrow (x)(Fx \leftrightarrow P)$
   (e) $(x)(\exists y)(Fx \leftrightarrow Gy) \leftrightarrow (\exists y)(x)(Fx \leftrightarrow Gy)$
   (f) $(x)(Fx \rightarrow (\exists y)Gy) \leftrightarrow (x)(\exists y)(Fx \rightarrow Gy)$
   (g) $(\exists x)((\exists x)Fx \rightarrow Fx)$
   (h) $(x)(Fx \leftrightarrow Gx) \leftrightarrow ((x)Fx \leftrightarrow (x)Gx)$
   (i) $(x)(Fx \leftrightarrow Gx) \rightarrow ((x)Fx \leftrightarrow (x)Gx)$
   (j) $(\exists x)(y)(Fx \rightarrow Gy) \leftrightarrow (\exists y)(x)(Fx \rightarrow Gy)$
8. Give an example of a theorem of logic of the form $(\exists\alpha)\phi$ which is such that the corresponding sentence $(\alpha)\phi$ is not a theorem of logic.
9. Find a formula ϕ such that ϕ contains free occurrences of the variables α and β, and all predicates in ϕ are of degree 1, and

$$(\exists\alpha)(\beta)\phi \leftrightarrow (\beta)(\exists\alpha)\phi$$

is a sentence but is not valid.

10. Symbolize the following arguments (taken from Lewis Carroll, *Symbolic Logic*, pp. 113ff.) relative to suitable interpretations, and in each case derive the symbolized conclusion from the symbolized premises, using any of the rules of this chapter.

(*a*)

(1) No terriers wander among the signs of the zodiac;
(2) Nothing that does not wander among the signs of the zodiac is a comet;
(3) Nothing but a terrier has a curly tail;
(4) Therefore, no comet has a curly tail.

(*b*)

(1) Nobody who really appreciates Beethoven fails to keep silence while the Moonlight Sonata is being played;
(2) Guinea pigs are hopelessly ignorant of music;
(3) No one who is hopelessly ignorant of music ever keeps silence while the Moonlight Sonata is being played;
(4) Therefore, guinea pigs never really appreciate Beethoven.

11. Do the same for the following arguments, from the same source:

(*a*)

(1) Animals are always mortally offended if I fail to notice them;
(2) The only animals that belong to me are in that field;
(3) No animal can guess a conundrum unless it has been properly trained in a Board-school;
(4) None of the animals in that field are badgers;
(5) When an animal is mortally offended, it always rushes about wildly and howls;

(6) I never notice any animal unless it belongs to me;
(7) No animal that has been properly trained in a Board-school ever rushes about wildly and howls;
(8) Therefore, no badger can guess a conundrum.

(*b*)

(1) The only animals in this house are cats;
(2) Every animal that loves to gaze at the moon is suitable for a pet;
(3) When I detest an animal, I avoid it;
(4) No animals are carnivorous unless they prowl at night;
(5) No cat fails to kill mice;
(6) No animals ever take to me, except what are in this house;
(7) Kangaroos are not suitable for pets;
(8) None but carnivora kill mice;
(9) I detest animals that do not take to me;
(10) Animals that prowl at night always love to gaze at the moon;
(11) Therefore, I always avoid a kangaroo.

8

SOME METATHEOREMS

That the rules of inference presented in Chapter 7 are worthy of the name cannot be made evident merely by giving a large number of particular derivations. We must show that in general, if a sentence ϕ is derivable by these rules from a set of sentences Γ, then ϕ is a consequence of Γ, and, conversely, that if a sentence ϕ is a consequence of a set of sentences Γ, then ϕ is derivable from Γ by the rules. The first of these results is established in section 2 of the present chapter, the second in section 3. Section 1 is devoted to the statement and proof of various other general principles ('metatheorems') characterizing the language $\mathfrak{L}$.

1. *Replacement, negation and duality, prenex normal form.* Instead of continuing to list particular instances of theorems of logic, let us endeavor to characterize this important class of sentences in a more general way. To this end we must bring in some further notation.

First of all, extending a device introduced on page 50, we shall say that, for any formula ϕ, distinct (i.e., such that no two are identical) variables $\alpha_1, \ldots, \alpha_n$ and individual symbols (not necessarily distinct) $\beta_1, \ldots, \beta_n$, $\phi\,^{\alpha_1 \ldots \alpha_n}_{\beta_1 \ldots \beta_n}$ is the result of replacing all free occurrences of $\alpha_1, \ldots, \alpha_n$ in ϕ by occurrences of $\beta_1, \ldots, \beta_n$, respectively. (If ϕ is not a formula, or $\alpha_1, \ldots, \alpha_n$ are not distinct variables, or $\beta_1, \ldots, \beta_n$ are not individual symbols, the notation is not defined).

Thus, for example, if ϕ = '$(x)(Fxyz \rightarrow (\exists z)Gzyua)$', α_1 = 'y', α_2 = 'u', β_1 = 'a', and β_2 = 'b', then the formula $\phi^{\alpha_1 \alpha_2}_{\beta_1 \beta_2}$ is '$(x)(Fxaz \rightarrow (\exists z)Gzaba)$'.

In formulating our generalizations we shall also make use of the concept of a 'closure' of a formula. It is defined in terms of the phrase 'string of quantifiers', which in turn is defined in the obvious way as follows:

An expression is a *string of quantifiers* if and only if it is either a quantifier standing alone or else is the result of prefixing a quantifier to a shorter string of quantifiers. (An expression is a *string of universal quantifiers* if and only if it is a string of quantifiers and every quantifier in it is universal; analogously for a *string of existential quantifiers.*)

A formula ϕ is said to be a *closure* of a formula ψ if and only if ϕ is a sentence and either $\phi = \psi$ or ϕ is the result of prefixing a string of universal quantifiers to ψ.

We further define the symbol '$\Vdash$' as follows: for any formula ϕ, $\Vdash\phi$ if and only if every closure of ϕ is a theorem of logic.

Examples: '$(x)Fx$' is a closure of 'Fx' and of itself; '$(x)(y)Fxy$' is a closure of 'Fxy', as well as of '$(y)Fxy$' and of itself; '$(x)Fa$' is a closure of 'Fa' and of itself; 'Fa' is a closure of itself. '$(\exists x)Fx$', on the other hand, is a closure of itself but not of 'Fx'; '$(x)(y)(z)Fxyzu$' is not a closure of anything. In general, the closures of any formula are obtained by prefixing enough universal quantifiers to convert it into a sentence. If it is already a sentence, no universal quantifiers need be prefixed; hence, every sentence is a closure of itself. All closures are sentences, but that of which a sentence is a closure may not be a sentence. Every formula has infinitely many closures, but of course a given sentence can be a closure of only finitely many formulas.

We are now in a position to state and to give informal proofs of a number of metatheoretic generalizations about the theorems of logic. In the following statements, unless otherwise noted, the range of 'ϕ', 'ψ', 'χ', 'ϕ'', 'ψ'', and 'χ'' consists of formulas; that of 'α', 'α'', 'α_1', etc. consists of variables, and that of 'β', 'β'', 'β_1', etc. consists of individual constants.

I. If $\phi^{\alpha_1 \ldots \alpha_n}_{\beta_1 \ldots \beta_n}$ is a theorem of logic (where $\beta_1, \ldots, \beta_n$ are distinct individual constants not occurring in ϕ), then $\Vdash\phi$.

Proof: Suppose that $\phi^{\alpha_1 \ldots \alpha_n}_{\beta_1 \ldots \beta_n}$ is a theorem of logic, in other words, that there is a derivation of this sentence from the empty set. To obtain a derivation of any given closure of ϕ we have only to add lines in which the required quantifiers are prefixed, applying UG at each step. Such application is possible since by hypothesis the individual constants $\beta_1, \ldots, \beta_n$ are distinct from one another and from all individual constants occurring in ϕ.

Note that the parenthetical proviso of I is necessary; '$Fa \vee -Fa$', for instance, is a theorem, but '$(x)(Fx \vee -Fa)$' is not.

II. If one closure of ϕ is a theorem of logic, then $\Vdash\phi$.

Proof: Assume that one closure of ϕ is a theorem of logic. By repeated application of US extend the derivation of this closure to obtain a derivation of a sentence $\phi^{\alpha_1 \ldots \alpha_n}_{\beta_1 \ldots \beta_n}$, where the variables $\alpha_1, \ldots, \alpha_n$ are distinct and include all free variables of ϕ, and $\beta_1, \ldots, \beta_n$ are distinct individual constants not occurring in ϕ. This sentence $\phi^{\alpha_1 \ldots \alpha_n}_{\beta_1 \ldots \beta_n}$ is a theorem of logic. Therefore, by I, $\Vdash \phi$.

III. If $\Vdash \phi$ and $\Vdash \phi \rightarrow \psi$, then $\Vdash \psi$.

Proof: Suppose that $\alpha_1, \ldots, \alpha_n$ are distinct variables and include all variables occurring free in $\phi \rightarrow \psi$. Then, by hypothesis, $(\alpha_1) \ldots (\alpha_n)\phi$ and $(\alpha_1) \ldots (\alpha_n)(\phi \rightarrow \psi)$ are theorems of logic. By repeated application of US, $\phi^{\alpha_1 \ldots \alpha_n}_{\beta_1 \ldots \beta_n}$ and $(\phi \rightarrow \psi)^{\alpha_1 \ldots \alpha_n}_{\beta_1 \ldots \beta_n}$ are theorems, where $\beta_1, \ldots, \beta_n$ are distinct individual constants not occurring in $\phi \rightarrow \psi$. Therefore, by rule T, $\psi^{\alpha_1 \ldots \alpha_n}_{\beta_1 \ldots \beta_n}$ is a theorem. By I, $\Vdash \psi$.

IV. (Generalizations of theorems 1–28, Chapter 7.)

1. $\Vdash (\alpha)(\alpha')\phi \leftrightarrow (\alpha')(\alpha)\phi$
2. $\Vdash (\exists\alpha)(\exists\alpha')\phi \leftrightarrow (\exists\alpha')(\exists\alpha)\phi$

   etc.

9. $\Vdash (\alpha)(\phi \,\&\, \psi) \leftrightarrow (\phi \,\&\, (\alpha)\psi)$ if α does not occur free in ϕ.

   etc.

18. $\Vdash (\alpha)(\exists\alpha')(\phi \,\&\, \psi) \leftrightarrow ((\alpha)\phi \,\&\, (\exists\alpha')\psi)$ if α does not occur free in ψ and α' does not occur free in ϕ.

   etc.

Proof: In each case the proof will involve constructing a schematic derivation analogous to the derivation of the corresponding theorem.

V. If $\Vdash \psi \leftrightarrow \psi'$, then $\Vdash -\psi \leftrightarrow -\psi'$

$\Vdash (\psi \,\&\, \chi) \leftrightarrow (\psi' \,\&\, \chi)$

$\Vdash (\chi \,\&\, \psi) \leftrightarrow (\chi \,\&\, \psi')$

$\Vdash (\psi \vee \chi) \leftrightarrow (\psi' \vee \chi)$

$\Vdash (\chi \vee \psi) \leftrightarrow (\chi \vee \psi')$

$\Vdash (\psi \rightarrow \chi) \leftrightarrow (\psi' \rightarrow \chi)$

$\Vdash (\chi \rightarrow \psi) \leftrightarrow (\chi \rightarrow \psi')$

$\Vdash (\psi \leftrightarrow \chi) \leftrightarrow (\psi' \leftrightarrow \chi)$

$\Vdash (\chi \leftrightarrow \psi) \leftrightarrow (\chi \leftrightarrow \psi')$

$\Vdash (\alpha)\psi \leftrightarrow (\alpha)\psi'$

$\Vdash (\exists\alpha)\psi \leftrightarrow (\exists\alpha)\psi'$

Proof: Since every tautology is a theorem of logic, we see by I that $\Vdash (\psi \leftrightarrow \psi') \rightarrow (-\psi \leftrightarrow -\psi')$. Hence, by III, the first part of the present metatheorem follows, and similarly the next eight parts. For the last two parts, assume that $\Vdash \psi \leftrightarrow \psi'$. By I and III, $\Vdash \psi \rightarrow \psi'$ and $\Vdash \psi' \rightarrow \psi$. Hence $\Vdash (\alpha)(\psi \rightarrow \psi')$ and $\Vdash (\alpha)(\psi' \rightarrow \psi)$, because every closure of $(\alpha)(\psi \rightarrow \psi')$ is a closure of $\psi \rightarrow \psi'$, and every closure of $(\alpha)(\psi' \rightarrow \psi)$ is a closure of $\psi' \rightarrow \psi$. By III and IV (theorems 6 and 15), $\Vdash (\alpha)\psi \rightarrow (\alpha)\psi'$, $\Vdash (\alpha)\psi' \rightarrow (\alpha)\psi$, $\Vdash (\exists\alpha)\psi \rightarrow (\exists\alpha)\psi'$, and $\Vdash (\exists\alpha)\psi' \rightarrow (\exists\alpha)\psi$. But, by I, $\Vdash ((\alpha)\psi \rightarrow (\alpha)\psi') \rightarrow (((\alpha)\psi' \rightarrow (\alpha)\psi) \rightarrow ((\alpha)\psi \leftrightarrow (\alpha)\psi'))$, and hence by III, $\Vdash (\alpha)\psi \leftrightarrow (\alpha)\psi'$. Similarly, $\Vdash (\exists\alpha)\psi \leftrightarrow (\exists\alpha)\psi'$.

VI. (Replacement). Suppose that ϕ' is like ϕ except for containing an occurrence of ψ' where ϕ contains an occurrence of ψ, and suppose that $\Vdash \psi \leftrightarrow \psi'$. Then $\Vdash \phi \leftrightarrow \phi'$, and $\Vdash \phi$ if and only if $\Vdash \phi'$.

Proof: Under the given supposition, ϕ can be built up from the given occurrence of ψ by means of the connectives and quantifiers, and ϕ' can be built up in an exactly corresponding way from the occurrence of ψ'; hence by repeated use of V we have $\Vdash \phi \leftrightarrow \phi'$. From this and I and III we have the further (and much weaker) conclusion that $\Vdash \phi$ if and only if $\Vdash \phi'$.

Let us define 'equivalence' among formulas as follows: for any formulas ϕ and ψ, ϕ is *equivalent* to ψ if and only if $\Vdash \phi \leftrightarrow \psi$. Thus metatheorem VI says that if in a formula ϕ we replace an occurrence of a formula ψ by an occurrence of a formula equivalent to ψ, the whole result is a formula equivalent to ϕ.

VII. (Rewriting of bound variables). If formulas $(\alpha)\phi$ and $(\alpha')\phi'$ are alike except that the former has occurrences of α where and only where the latter has occurrences of α', then they are equivalent. Similarly for $(\exists\alpha)\phi$ and $(\exists\alpha')\phi'$.

Proof: Assume the hypothesis. Let $\alpha_1, \ldots, \alpha_n$ be a list (without duplications) of all variables that occur free in $(\alpha)\phi$ (and hence in $(\alpha')\phi'$), and let $\beta_1, \ldots, \beta_n$ be distinct individual constants not occurring in ϕ (and hence again not in ϕ'). By an obvious derivation we see that $((\alpha)\phi \leftrightarrow (\alpha')\phi')\,{}^{\alpha_1 \ldots \alpha_n}_{\beta_1 \ldots \beta_n}$ is a theorem of logic. By I, $(\alpha)\phi$ and $(\alpha')\phi'$ are equivalent. The argument for the existential quantifier is similar.

VIII. (Negation theorem). Suppose that a formula ϕ contains no occurrences of '$\rightarrow$' and '$\leftrightarrow$' and that ϕ' is obtained from ϕ by exchanging '&' and 'v', exchanging universal and the corresponding existential quantifiers, and by replacing atomic formulas by their negations. Then ϕ' is equivalent to $-\phi$.

Proof: We proceed by showing (a) that VIII holds for atomic formulas, and (b) that it holds for all negations, conjunctions, disjunctions, and generalizations of the formulas for which it holds. This implies that it holds for all formulas not containing '$\rightarrow$' or '$\leftrightarrow$'.

(a) If ϕ is atomic, then ϕ' *is* $-\phi$, and hence ϕ' is equivalent to $-\phi$.

(b) (i) Suppose that $\phi = -\psi_1$, and that VIII holds of ψ_1. Then ψ_1' is equivalent to $-\psi_1$, i.e., to ϕ. Therefore, $-\psi_1'$, i.e., ϕ', is equivalent to $-\phi$.

(ii) Suppose that $\phi = \psi_1 \,\&\, \psi_2$ and that VIII holds of ψ_1 and of ψ_2. Then $-\phi$ is equivalent to $-\psi_1 \vee -\psi_2$, which, by hypothesis and VI, is equivalent to $\psi_1' \vee \psi_2'$, i.e., to ϕ'.

(iii) Suppose that $\phi = \psi_1 \vee \psi_2$ and that VIII holds of ψ_1 and of ψ_2. Then $-\phi$ is equivalent to $-\psi_1 \,\&\, -\psi_2$, which, by hypothesis and VI, is equivalent to $\psi_1' \,\&\, \psi_2'$, i.e., to ϕ'.

(iv) Suppose that $\phi = (\alpha)\psi_1$, and that VIII holds of ψ_1. Then $-\phi$ is equivalent to $(\exists\alpha)-\psi_1$, which, by hypothesis and VI, is equivalent to $(\exists\alpha)\psi_1'$, i.e., to ϕ'.

(v) Suppose that $\phi = (\exists\alpha)\psi_1$, and that VIII holds of ψ_1. Then $-\phi$ is equivalent to $(\alpha)-\psi_1$, which, by hypothesis and VI, is equivalent to $(\alpha)\psi_1'$, i.e., to ϕ'.

Example 1. Apply VIII to find a formula equivalent to the negation of

$$(x)(\exists y)-(Fxy \vee (\exists z)(Gzy \,\&\, (Hxz \vee -Fxz))).$$

The desired formula is

$$(\exists x)(y)-(-Fxy \,\&\, (z)(-Gzy \vee (-Hxz \,\&\, --Fxz))).$$

Example 2. By repeated application of VIII and VI to parts of the following formula (dropping double negations where possible), find an equivalent formula in which the negation sign stands only ahead of predicate letters.

$$-(y)(z)-(\exists x)-(-Fxz \vee (w)-(Hxzw \,\&\, -Hxzx)).$$

Successive steps on the way to the desired formula are:

$$(\exists y)(\exists z)-(x)-(Fxz \,\&\, (\exists w)-(-Hxzw \vee Hxzx))$$
$$(\exists y)(\exists z)(\exists x)-(-Fxz \vee (w)-(Hxzw \,\&\, -Hxzx))$$
$$(\exists y)(\exists z)(\exists x)(Fxz \,\&\, (\exists w)-(-Hxzw \vee Hxzx))$$
$$(\exists y)(\exists z)(\exists x)(Fxz \,\&\, (\exists w)(Hxzw \,\&\, -Hxzx))$$

For our next metatheorem we need another definition: a formula ϕ is in *prenex normal form* if and only if ϕ is either quantifier-free or consists of a string of quantifiers followed by a quantifier-free formula.

IX. (Prenex normal form). For any formula ϕ there is an equivalent formula ψ that is in prenex normal form.

Proof: We first describe a procedure (called 'reduction to prenex normal form') for obtaining an equivalent prenex formula ψ corresponding to any given formula ϕ. Given ϕ, we obtain ψ in four stages as follows.

1. First we eliminate all occurrences of '$\rightarrow$' and '$\leftrightarrow$' from ϕ by systematically replacing (VI) all parts of the form $\psi_1 \rightarrow \psi_2$ by the corresponding formulas of the form $-\psi_1 \vee \psi_2$, and by replacing all parts of the form $\psi_1 \leftrightarrow \psi_2$ by the corresponding formulas of the form $(\psi_1 \,\&\, \psi_2) \vee (-\psi_1 \,\&\, -\psi_2)$.

2. To the result of this we apply the procedure indicated in example 2 to VIII above, 'bringing in' negation signs until they stand only ahead of predicate letters.

3. Next we rewrite the quantifiers and bound variables until no two quantifier-occurrences contain the same variable and no variable occurs both free and bound.

4. Finally we place all the quantifiers at the beginning of the formula, in the order in which they occur.

To illustrate this procedure, let us reduce the following formula to prenex normal form:

$$(x)Fx \leftrightarrow (\exists x)Gx.$$

Step 1 yields

$$((x)Fx \,\&\, (\exists x)Gx) \vee (-(x)Fx \,\&\, -(\exists x)Gx).$$

Bringing in the negation signs, as directed in step 2, we have

$$((x)Fx \,\&\, (\exists x)Gx) \vee ((\exists x)-Fx \,\&\, (x)-Gx).$$

Rewriting the quantifiers according to step 3:

$$((x)Fx \,\&\, (\exists y)Gy) \vee ((\exists z)-Fz \,\&\, (u)-Gu).$$

Bringing out the quantifiers in the order in which they occur, we get the desired formula in prenex normal form:

$$(x)(\exists y)(\exists z)(u)((Fx \,\&\, Gy) \vee (-Fz \,\&\, -Gu)).$$

The proof of this metatheorem consists in establishing that the various replacements described in steps 1–4 always lead to formulas equivalent to those to which they are applied. That this is true of the replacements involved in step 1 follows from the fact that every tautology is a theorem of logic, and so, for any formulas ϕ and ψ we have, by I,

$$\Vdash (\phi \rightarrow \psi) \leftrightarrow (-\phi \vee \psi) \quad \text{and}$$
$$\Vdash (\phi \leftrightarrow \psi) \leftrightarrow ((\phi \,\&\, \psi) \vee (-\phi \,\&\, -\psi)).$$

The replacements of step 2 are justified by VI and VIII (in view of VI and the fact that $\Vdash \phi \leftrightarrow --\phi$, double negations may be dropped). Step 3 is justified by the proof of VII. Finally, the replacements involved in step 4 are justified by IV and VI (theorems 9, 10, 12, 13, and the corresponding theorems with conjunctions and disjunctions commuted).

For a second example of reduction to prenex normal form we apply the procedure to

$$((x)(\exists y)(Fx \rightarrow Gy) \,\&\, (\exists x)Fx) \rightarrow (\exists y)Gy.$$

Successive steps on the way to the desired formula are:

$$-((x)(\exists y)(-Fx \vee Gy) \,\&\, (\exists x)Fx) \vee (\exists y)Gy$$
$$((\exists x)(y)(Fx \,\&\, -Gy) \vee (x)-Fx) \vee (\exists y)Gy$$
$$((\exists x)(y)(Fx \,\&\, -Gy) \vee (z)-Fz) \vee (\exists u)Gu$$
$$(\exists x)(y)(z)(\exists u)(((Fx \,\&\, -Gy) \vee -Fz) \vee Gu).$$

The prenex normal form metatheorem has proved to be a very useful tool in a number of metatheoretic investigations. Instead of considering all possible formulas of $\mathfrak{L}$ one can restrict oneself, for many purposes, to formulas in prenex normal form. Thus, in his original proof that every valid sentence is a theorem of logic, K. Gödel needed only to show that every valid sentence in prenex normal form is a theorem. Also, many of the results that have been obtained in connection with the so-called decision problem utilize the metatheorem at hand. Although, as stated near the beginning of Chapter 7, there can be no step-by-step procedure for deciding whether or not an arbitrarily given sentence of $\mathfrak{L}$ is valid, such procedures have been found for certain restricted classes of sentences. The most interesting of these classes are characterized in terms of the prenex normal forms of their members. For instance, it has been shown that if a sentence ϕ has a prenex normal form in the prefix of which no existential quantifier precedes any universal quantifier, then the validity of ϕ is decidable. For a list of similar results, see Alonzo Church, *Introduction to Mathematical Logic,* section 46.

X. If $\Vdash\phi$ and if ψ is the result of replacing all atomic formulas in ϕ by their negations, then $\Vdash\psi$.

Proof: We leave to the reader the task of checking that any derivation will remain a derivation when atomic formulas are replaced by their negations throughout.

XI. (Duality). If $\Vdash\phi \leftrightarrow \psi$ and neither '$\rightarrow$' nor '$\leftrightarrow$' occurs in ϕ or ψ, and if ϕ^* and ψ^* are obtained from ϕ and ψ, respectively, by exchanging '&' and 'v', and universal and the corresponding existential quantifiers, then $\Vdash\phi^* \leftrightarrow \psi^*$; similarly, if $\Vdash\phi \rightarrow \psi$, then $\Vdash\psi^* \rightarrow \phi^*$.

Proof: Assume the hypothesis of XI. Then $\Vdash -\phi \leftrightarrow -\psi$, and by VIII and VI, $\Vdash \phi' \leftrightarrow \psi'$. By X, and dropping double negations, $\phi^* \leftrightarrow \psi^*$ is a theorem. The proof of the last clause is analogous.

Example 1. Since

$$(x)(Fx \,\&\, Gx) \leftrightarrow ((x)Fx \,\&\, (x)Gx)$$

is a theorem of logic, the duality law informs us that

$$(\exists x)(Fx \vee Gx) \leftrightarrow ((\exists x)Fx \vee (\exists x)Gx)$$

is also a theorem of logic.

Example 2. The dual of the theorem

$$(\exists x)((y)Fxy \vee (y)-Fxy) \leftrightarrow (\exists x)(y)(z)(Fxy \vee -Fxz)$$

is the theorem

$$(x)((\exists y)Fxy \,\&\, (\exists y)-Fxy) \leftrightarrow (x)(\exists y)(\exists z)(Fxy \,\&\, -Fxz).$$

2. *Soundness and consistency*. To say that a system of inference rules is *sound* is to say that any conclusion derived by their use will be a consequence of the premises from which it is obtained. Thus, to show that the present system, consisting of P, T, C, US, UG, and E, is sound, we must show that, for any sentence ϕ and set of sentences Γ, if ϕ is derivable from Γ, then ϕ is a consequence of Γ. A system of rules is called *consistent* if there is no sentence ϕ such that both ϕ and $-\phi$ are derivable from Λ. Obviously, if rules are sound they are consistent, for no sentence and its

negation are both consequences of Λ. On the other hand, consistency does not guarantee soundness. It is easy to think of a rule that would be consistent but not sound—e.g., a rule that would allow us to derive '*P*' (and no other sentence) from every set of sentences.

Let us now establish the soundness of our rules. Roughly, the argument amounts to this: rule P allows us to enter premises (as consequences of themselves), and each of the other five rules allows us to write down only such sentences as are consequences of sentences appearing on earlier lines. Thus, by repeated application of the rules we get only sentences that are consequences of the premises.

Let us state this argument more carefully. We assert that (1) any sentence appearing on the first line of a derivation is a consequence of the premises of that line, and (2) any sentence appearing on a later line is a consequence of its premises if all sentences appearing on earlier lines are consequences of theirs.

As concerns (1): if ϕ appears on the first line, ϕ was entered by rule P (in which case it is its own only premise) or by rule T (in which case it is a tautology and hence a consequence of the empty set of sentences).

Concerning (2) we consider the rules one at a time.

(i) If ϕ was entered on a line by rule P, it is obviously a consequence of the premises of that line.

(ii) If ϕ was entered on a line by rule T, it is a tautological consequence of a set Γ of sentences that appear on earlier lines. Thus ϕ is a consequence of Γ (see page 90). But all the sentences of Γ are, by hypothesis, consequences of the premises of the lines on which they appear. Call the totality of these premises Δ. Then ϕ is a consequence of Δ (page 65, item 1).

(iii) If ϕ was entered on a line by rule C, then $\phi = (\psi \rightarrow \chi)$, where χ appears on an earlier line. By hypothesis, χ is a consequence of the premises of its line, which may include ψ. Therefore, $(\psi \rightarrow \chi)$, i.e., ϕ, is a consequence of these premises excluding ψ (page 66, item 6).

(iv) If ϕ was entered on a line by rule US, then $\phi = \psi\, \alpha/\beta$ and $(\alpha)\psi$ appears on an earlier line. By hypothesis, $(\alpha)\psi$ on that line is a consequence of its premises. But $\psi\ \alpha/\beta$ is a consequence of $(\alpha)\psi$ (page 66, item 9). Therefore, $\psi\ \alpha/\beta$, i.e., ϕ, is a consequence of the premises of the line on which $(\alpha)\psi$ appears (page 65, item 1).

(v) If ϕ was entered on a line by rule UG, then $\phi = (\alpha)\psi$ and $\psi\, \alpha/\beta$ appears on an earlier line, with premises Γ, where β occurs neither in ψ nor in any member of Γ. By hypothesis, $\psi\, \alpha/\beta$ is a consequence of Γ. Therefore, $(\alpha)\psi$, i.e., ϕ, is also a consequence of Γ (page 66, item 10).

(vi) If ϕ was entered on a line by rule E, then either $\phi = (\exists\alpha)\psi$ and $-(\alpha)-\psi$ appears on an earlier line, or $\phi = -(\alpha)-\psi$ and $(\exists\alpha)\psi$ appears on an earlier line. In either case, since the sentence appearing on the earlier line is by hypothesis a consequence of the premises of that line, and ϕ is a consequence of the sentence appearing on the earlier line (page 66, item 11), ϕ is a consequence of the premises of the earlier line (page 65, item 1).

Thus, any sentence appearing on a line of a derivation is a consequence of the premises of that line; in other words, if a sentence ϕ is derivable from a set of sentences Γ, then ϕ is a consequence of Γ. On the basis of this fact, some advice can be given to anyone attempting to construct derivations in accord with our rules: if at any stage you find that you have entered a sentence which does not seem intuitively to be a consequence of the premises of that line, check the derivation, for either your intuition is incorrect or there has been an error in the application of some rule. Of course it does not follow that if what you write *is* a consequence of its premises, then the rules *have* been followed; we know only that if it is *not* a consequence, the rules have *not* been followed.

The consistency of our rules may also be established by a purely syntactical argument, i.e., by one that does not utilize the notion of interpretation or any similar notion.* It may be of some interest to look at that argument.

For any sentence ϕ we define the *transform* of ϕ ($\mathfrak{T}(\phi)$) as that SC sentence which is the result of deleting all individual symbols, quantifiers, and predicate superscripts from ϕ. For example, the transform of

$$(F_1 \rightarrow (x)(F_1^2 xa \vee (\exists y)(G_2^3 xay \,\&\, F_1^1 x)))$$

is

$$(F_1 \rightarrow (F_1 \vee (G_2 \,\&\, F_1))).$$

Now if, in a correct derivation, we replace each sentence by its transform, leaving premise-numbers unchanged, we obtain another derivation in which each line can be justified by one of the rules P, T, or C. Thus, if a pair of sentences ϕ and $-\phi$ were derivable from Λ by our rules, $\mathfrak{T}(\phi)$ and $\mathfrak{T}(-\phi)$ would be tautologies; but $\mathfrak{T}(-\phi)$ will be the negation of $\mathfrak{T}(\phi)$, and it is impossible for any sentence and its negation both to be tautologous. Therefore, no pair of sentences ϕ and $-\phi$ are derivable from Λ by our rules.

To see that when a derivation is transformed in the way indicated every line of the resulting derivation may be justified by rules P, T, or C, one considers in turn each of the rules P, T, C, US, UG, and E. If a line of the original derivation was entered in accord with rule P, the corresponding line of the new derivation will likewise be justified by P. The same holds for rules T and C. Inferences by US and UG become simple repetitions under the transformation we are considering, and rule E always takes us from ϕ to $--\phi$ or conversely.

* In view of the (syntactic) truth-table test for tautologousness, however, it is cricket to use the notion of 'tautology', which, as some readers have pointed out, we defined semantically.

3. *Completeness.* (This section may be omitted without loss of continuity). To say that a system of inference rules is *complete* is to say that by its use one can derive, from any given set of sentences, any consequence of that set. Thus, to show that the present system, consisting of P, T, C, US, UG, and E, is complete, we must show that, for any sentence ϕ and set of sentences Γ, if ϕ is a consequence of Γ, then ϕ is derivable from Γ. The following proof is due in its essentials to Professor Leon Henkin.

Henkin's proof is formulated in terms of certain notions that possess interest even independently of their use in the present connection.

We say that a set of sentences Γ is *consistent with respect to derivability* (*d-consistent*) if and only if the sentence '$P \& -P$' is not derivable from Γ.

Note that derivability is closely related to d-consistency: for any sentence ϕ and set of sentences Γ, ϕ is derivable from Γ if and only if $\Gamma \cup \{-\phi\}$ is not d-consistent. For suppose that ϕ is derivable from Γ. Then there is a derivation in which ϕ appears on the last line, say the i-th, and all premises of that line belong to Γ. By adding two lines as follows:

$\{n_1, \ldots, n_p\}$	(i)	ϕ	
$\{i+1\}$	$(i+1)$	$-\phi$	P
$\{n_1, \ldots, n_p, i+1\}$	$(i+2)$	$P \& -P$	$i, i+1$ T

we obtain a derivation of '$P \& -P$' from $\Gamma \cup \{-\phi\}$, so that $\Gamma \cup \{-\phi\}$ is not d-consistent. On the other hand, suppose that $\Gamma \cup \{-\phi\}$ is not d-consistent. Then '$P \& -P$' is derivable from $\Gamma \cup \{-\phi\}$. If we extend this derivation by an obvious application of rule C, we shall have a derivation of the sentence

$$-\phi \rightarrow (P \& -P)$$

from Γ, and, adding one more line, this time by rule T, we obtain a derivation of ϕ from Γ.

Also worth noting is the fact that for any set of sentences Γ, Γ is d-consistent if and only if there is at least one sentence ϕ of $\mathfrak{L}$ that is not derivable from Γ. In other words, '$P \& -P$' is derivable from a given set of sentences if and only if *every* sentence is so derivable.

Next, we define a set of sentences Γ as *maximal d-consistent* if and only if Γ is d-consistent and is not properly included in any d-consistent set Δ. Thus Γ is maximal d-consistent if and only if Γ is d-consistent but loses its d-consistency if we add to it any sentence not already a member.

Now let us note some of the properties of sets that are maximal d-consistent. Let Δ be any such set, and let ϕ be any sentence of $\mathfrak{L}$. Then we have

(1) $\phi \in \Delta$ iff $-\phi \notin \Delta$, and
(2) $\phi \in \Delta$ iff ϕ is derivable from Δ.*

*'Iff' for 'if and only if'.

Proof of (1): Suppose that both ϕ and $-\phi$ belong to Δ. Then both are derivable from Δ, and thus '$P \& -P$' is derivable from Δ, contrary to the hypothesis that Δ is d-consistent. On the other hand, suppose that neither ϕ nor $-\phi$ belongs to Δ. Since Δ is maximal d-consistent, this means that both $\Delta \cup \{\phi\}$ and $\Delta \cup \{-\phi\}$ are not d-consistent. Again, therefore, both ϕ and $-\phi$ are derivable from Δ, contrary to hypothesis. Thus, exactly one of the pair ϕ, $-\phi$ belongs to Δ, as is asserted by (1). Proof of (2): If $\phi \in \Delta$, then obviously ϕ is derivable from Δ. Suppose $\phi \notin \Delta$. Then $-\phi \in \Delta$, and hence $-\phi$ is derivable from Δ. Thus, since Δ is d-consistent, ϕ is not derivable from Δ.

From (1) and (2) it is easy to get the following consequences, where Δ is a maximal d-consistent set of sentences and ϕ, ψ are any sentences:

(3) $(\phi \vee \psi) \in \Delta$ iff $\phi \in \Delta$ or $\psi \in \Delta$.

(4) $(\phi \& \psi) \in \Delta$ iff $\phi \in \Delta$ and $\psi \in \Delta$.

(5) $(\phi \rightarrow \psi) \in \Delta$ iff $\phi \notin \Delta$ or $\psi \in \Delta$, or both.

(6) $(\phi \leftrightarrow \psi) \in \Delta$ iff $\phi, \psi \in \Delta$ or $\phi, \psi \notin \Delta$.

Let us say next that a set of sentences Γ is *ω-complete* if and only if it satisfies the following condition: for every formula ϕ and variable α, if $(\exists\alpha)\phi$ belongs to Γ, then there is an individual constant β such that $\phi\, \alpha/\beta$ also belongs to Γ. In other words, an ω-complete set of sentences never contains an existential sentence without also containing a sentence from which it can be derived by one application of rule EG.

When a set of sentences Δ is both maximal d-consistent and ω-complete it has, of course, all the properties of maximal d-consistent sets (including (1)–(6) above) and in addition the following two:

(7) $(\alpha)\phi \in \Delta$ iff for every individual constant β, $\phi\, \alpha/\beta \in \Delta$.

(8) $(\exists\alpha)\phi \in \Delta$ iff for some individual constant β, $\phi\, \alpha/\beta \in \Delta$.

Proof of (7): The implication from left to right is obvious, in view of (2) and the rule US. Suppose now that $\phi\, \alpha/\beta \in \Delta$ for every individual constant β, but $(\alpha)\phi \notin \Delta$. Then, by (1), $-(\alpha)\phi \in \Delta$. By (2), $(\exists\alpha)-\phi \in \Delta$. As Δ is ω-complete, $-\phi\, \alpha/\gamma \in \Delta$, for some individual constant γ. But since $\phi\, \alpha/\gamma$ is also a member of Δ, Δ is not d-consistent, contrary to hypothesis. The proof of (8) is analogous.

Next we state the principal lemma for the completeness result.

I. For any set of sentences Γ, if Γ is d-consistent, then it is consistent.

Once this lemma is available, we can establish completeness as follows. Let ϕ be a sentence and Γ a set of sentences, and suppose that ϕ is a consequence of Γ. Then $\Gamma \cup \{-\phi\}$ is not consistent. Hence, by I, $\Gamma \cup \{-\phi\}$ is not d-consistent, and therefore ϕ is derivable from Γ. Thus, if ϕ is a consequence of Γ, ϕ is derivable from Γ; therefore, our system of rules is complete.

It is convenient to prove first a special case of I:

I′. For any set of sentences Γ, if Γ is d-consistent and all indices of individual constants occurring in the sentences of Γ are even, then Γ is consistent.

The advantage of the restriction about indices, which may at first strike the reader as rather odd, is that it guarantees that there are infinitely many individual constants not occurring in the sentences of Γ; this, as is soon to become evident, will be useful.

Clearly I follows from I′. For, given a set of sentences Γ, by doubling the indices of all individual constants occurring therein we can form a set Γ^* of the type described in I′. As this amounts only to replacing distinct constants by distinct constants, Γ^* will be consistent if and only if Γ is consistent, and Γ^* will be d-consistent if and only if Γ is d-consistent.

We shall prove I′ by means of two further lemmas, from which it follows immediately.

II. For any set of sentences Γ, if Γ satisfies the hypothesis of I′, then there is a set of sentences Δ that includes Γ and is maximal d-consistent and ω-complete.

III. Every maximal d-consistent, ω-complete set of sentences is consistent.

Proof of II: Suppose that Γ satisfies the hypothesis of I′.

1) We make use of the fact, here assumed without proof, that all the sentences of $\mathfrak{L}$ can be arranged in an infinite list

$$\phi_1, \phi_2, \phi_3, \ldots, \phi_n, \ldots$$

with the following properties:

(a) each sentence of $\mathfrak{L}$ occurs at least once in the list;

(b) for each sentence of the form $(\exists\alpha)\phi$, there is at least one i such that $\phi_i = (\exists\alpha)\phi$ and $\phi_{i+1} = \phi\ \alpha/\beta$, where β is a 'new' individual constant (i.e., not appearing in any of the sentences $\phi_1, \phi_2, \ldots, \phi_i$ or in any sentence of Γ).

2) Now, with respect to this list, we construct an infinite sequence of sets

$$\Delta_0, \Delta_1, \Delta_2, \ldots, \Delta_n, \ldots$$

as follows. As Δ_0 we take Γ, i.e.,

$$\Delta_0 = \Gamma.$$

We form Δ_1 by adding the sentence ϕ_1 to Δ_0 if $\Delta_0 \cup \{\phi_1\}$ is d-consistent; otherwise we let $\Delta_1 = \Delta_0$. In other words,

$$\Delta_1 = \begin{cases} \Delta_0 \cup \{\phi_1\} & \text{if this union is d-consistent,} \\ \Delta_0 & \text{otherwise.} \end{cases}$$

Passing then to the next sentence, ϕ_2, we add it to Δ_1 to form Δ_2 if the result of such addition is d-consistent; otherwise we let $\Delta_2 = \Delta_1$. That is,

$$\Delta_2 = \begin{cases} \Delta_1 \cup \{\phi_2\} & \text{if this union is d-consistent,} \\ \Delta_1 & \text{otherwise.} \end{cases}$$

In general, for each positive integer n we set

$$\Delta_n = \begin{cases} \Delta_{n-1} \cup \{\phi_n\} & \text{if this union is d-consistent,} \\ \Delta_{n-1} & \text{otherwise.} \end{cases}$$

And now let Δ be the union of all the infinitely many sets Δ_i. Thus, a sentence ϕ is an element of Δ if and only if it is an element of at least one of the sets $\Delta_0, \Delta_1, \ldots, \Delta_n, \ldots$. We proceed to show that Δ, which obviously includes Γ, is maximal d-consistent and ω-complete.

3) Δ is d-consistent. (i) Note that each Δ_i is d-consistent, by construction, since Γ is d-consistent. (ii) Suppose that '$P \,\&\, -P$' is derivable from Δ. Then '$P \,\&\, -P$' is derivable from a finite subset Δ' of Δ. (See p. 113, remark following definition of 'derivable from'.) But $\Delta' \subset \Delta_j$, for some j. (This follows from the fact that any finite subset of a union of nested sets is a subset of at least one of the nested sets.) So '$P \,\&\, -P$' is derivable from Δ_j, contrary to (i).

4) Δ is maximal d-consistent. For, suppose that a sentence ϕ is not a member of Δ. Now $\phi = \phi_i$, for some i. Since $\phi \notin \Delta, \phi \notin \Delta_i$. Therefore $\Delta_{i-1} \cup \{\phi\}$ is not d-consistent, and so $\Delta \cup \{\phi\}$ is not d-consistent.

5) Δ is ω-complete. Suppose that $(\exists\alpha)\phi \in \Delta$. By 1 (b) we know that for some i, $(\exists\alpha)\phi = \phi_i$ and $\phi\, \alpha/\beta = \phi_{i+1}$, where β occurs neither in Γ nor in any ϕ_j, $1 \leq j \leq i$. By the construction of Δ, $(\exists\alpha)\phi \in \Delta_i$. We see further that $\phi_{i+1} \in \Delta_{i+1}$ (and thus $\phi\, \alpha/\beta \in \Delta$). For, suppose the contrary; then $\Delta_i \cup \{\phi\, \alpha/\beta\}$ is not d-consistent, and consequently $-\phi\, \alpha/\beta$ is derivable from Δ_i. Since β occurs neither in Δ_i nor in ϕ, $(\alpha)-\phi$ is likewise derivable from Δ_i. Thus, both $(\alpha)-\phi$ and $-(\alpha)-\phi$ are derivable from Δ_i, which contradicts the d-consistency of Δ_i. Therefore, if $(\exists\alpha)\phi \in \Delta$, $\phi\, \alpha/\beta \in \Delta$.

Proof of III: Let Δ be a maximal d-consistent, ω-complete set of sentences. We specify an interpretation $\mathfrak{I}$ as follows. Let the domain be the set of all individual constants of $\mathfrak{L}$. With each individual constant of $\mathfrak{L}$ let $\mathfrak{I}$ associate as denotation that constant itself. With each sentential letter ϕ of $\mathfrak{L}$ let $\mathfrak{I}$ associate the truth-value T if $\phi \in \Delta$, otherwise F. With each n-ary predicate θ of $\mathfrak{L}$ let $\mathfrak{I}$ associate the set of those n-tuples of individual constants $\langle\gamma_1, \ldots, \gamma_n\rangle$ which are such that the result of writing θ followed by the string $\gamma_1\gamma_2 \ldots \gamma_n$ is an atomic sentence belonging to Δ.

We now show that, for every sentence ϕ of $\mathfrak{L}$, ϕ is true under this interpretation $\mathfrak{I}$ ('$\mathfrak{I}$-true') if and only if $\phi \in \Delta$. Let us define the *index* of a formula as the number of occurrences of quantifiers and/or connectives in the formula. Thus atomic sentences are of index 0. By the construction of $\mathfrak{I}$, our assertion holds obviously for atomic sentences. Suppose that there is at least one sentence of which the assertion does not hold (we shall deduce an absurdity from this supposition), and let ϕ be such a sentence of lowest index. In other words, it is not the case that ϕ is true under $\mathfrak{I}$ iff $\phi \in \Delta$, but, for every sentence ψ of index lower than that of ϕ, it *is* the case that ψ is true under $\mathfrak{I}$ iff $\psi \in \Delta$. Let n be the index of ϕ. Then $n > 0$, as noted above. So we have one of the following seven cases:

a) $\phi = -\psi$, where the index of ψ is less than n. But

$-\psi$ is $\mathfrak{I}$-true iff ψ is not $\mathfrak{I}$-true
iff $\psi \notin \Delta$ (since by hypothesis our assertion holds of ψ)
iff $-\psi \in \Delta$ (page 142, item (1))

Therefore, this case must be excluded.

b) $\phi = (\psi \vee \chi)$, where the indices of ψ, χ are less than n. But

$(\psi \vee \chi)$ is $\mathfrak{I}$-true iff ψ is $\mathfrak{I}$-true or χ is $\mathfrak{I}$-true
iff $\psi \in \Delta$ or $\chi \in \Delta$ (since by hypothesis our assertion holds of ψ and χ)
iff $(\psi \vee \chi) \in \Delta$ (page 143, item (3))

Therefore, this case also is excluded.

c) $\phi = (\psi \,\&\, \chi)$, where, etc. The argument is analogous, except that item (4) on page 143 is the relevant one.

d) $\phi = (\psi \rightarrow \chi)$, where, etc. The argument is analogous, except that item (5) on page 143 is the relevant one.

e) $\phi = (\psi \leftrightarrow \chi)$, where, etc. The argument is analogous, except that item (6) on page 143 is the relevant one.

f) $\phi = (\alpha)\psi$, where ψ is of index less than n. (Note too that for every individual constant β, the index of $\psi\, \alpha/\beta$ is the same as that of ψ). Now since every element of the domain of $\mathfrak{I}$ is assigned by $\mathfrak{I}$ to at least one individual constant, viz., to itself, we have (page 66, item 19)

$(\alpha)\psi$ is $\mathfrak{I}$-true iff for every individual constant β, $\psi\, \alpha/\beta$ is $\mathfrak{I}$-true
iff for every individual constant β, $\psi\, \alpha/\beta \in \Delta$ (by hypothesis)
iff $(\alpha)\psi \in \Delta$ (page 143, item (7))

Therefore, this case must also be excluded.

g) $\phi = (\exists\alpha)\psi$. The argument is similar to that in f above. Thus we have reached the absurdity that ϕ must fall under one of the cases a–g and yet cannot fall under any of them. This completes the proof of III, and thus, as explained earlier, we have I′, I, and the principal result.

It will be noted that in establishing III above we actually proved the following somewhat stronger result: every maximal d-consistent ω-complete set of sentences is satisfiable by an interpretation having a denumerably infinite domain (i.e., having a domain equinumerous with the positive integers). Thus in our proof of I we have shown that for any set of sentences Γ, if Γ is d-consistent, then it is satisfiable by an interpretation having a denumerably infinite domain. Since, in view of the soundness of our rules, if a set Γ is consistent, then it is d-consistent, we have the following metatheorem (the Löwenheim-Skolem theorem): if Γ is a consistent set of sentences, then Γ is satisfiable by an interpretation having a denumerably infinite domain.

We have seen earlier that one can easily find a consistent set of sentences Γ such that if all members of Γ are true under an interpretation $\mathfrak{I}$, then the domain of $\mathfrak{I}$ must contain at least two elements. For instance, the set {'*Fa*', '$-$*Fb*'} has this property. Similarly, the elements of {'*Fa*', '$-$*Fb*', '$-$*Fc*', '*Gb*', '$-$*Gc*'} can be simultaneously true under $\mathfrak{I}$ only if the domain of $\mathfrak{I}$ contains at least three elements, and the analogous procedure works for any positive integer *n*. Further, the set consisting of the three sentences given in exercise 11 (c), Chapter 4, is satisfiable only by interpretations having at least denumerably infinite domains. The question thus arises whether one could find a consistent set of sentences that is satisfiable only by interpretations having non-denumerably infinite domains. In view of the Löwenheim-Skolem theorem, the answer is negative. Thus if one attempts to characterize a mathematical structure by means of a set of axioms formulated in the first order predicate calculus, one is in a certain sense doomed to failure if that structure involves a non-denumerable infinity of elements.

4. *A proof procedure for valid sentences.* Although, as was stated at the beginning of Chapter 7, there can be no step-by-step procedure for deciding whether or not a given sentence of $\mathfrak{L}$ is valid, there *are* procedures which, *if* a given sentence *is* valid, will generate a proof of that sentence. If such a procedure, applied to a given sentence, produces a proof, we know of course that the sentence is valid; but if after any finite number of steps it has not yet produced a proof, both possibilities are still open—the sentence may be valid and we have not gone far enough to reach a proof, or it may be invalid and hence incapable of proof. Thus a proof procedure for valid sentences is one thing, and a decision procedure for validity is another; the former we shall now illustrate; the latter is impossible.

It is convenient to restrict ourselves to valid sentences that are in prenex normal form. This restriction, though convenient, is obviously inessential. For, given an arbitrary valid sentence ϕ, we can reduce ϕ to a sentence ψ in prenex normal form, which will also be valid; then we can use our procedure to generate a proof of ψ; and, finally, by reversing the steps of the reduction, extend this proof to a proof of ϕ. It is true that in most cases the actual derivation of a sentence and its prenex normal form from one another would be very long, but it would present no difficulties in principle.

Accordingly, let ϕ be a valid sentence in prenex normal form. Our procedure for generating a proof of ϕ is in outline as follows:

1) Enter $-\phi$ on the first line as a premise.

2) Derive a prenex normal form ψ from $-\phi$.

3) Now construct a sequence of lines, endeavoring to satisfy the following two conditions, until a truth-functional inconsistency appears:

(a) Whenever any universal generalization $(\alpha)\theta$ appears on a line, particular instances $\theta\, \alpha/\beta$ (for all individual constants β occurring in the sequence, and in any case for at least one individual constant β) shall appear on later lines, inferred by US;

(b) Whenever any existential generalization $(\exists\alpha)\theta$ appears on a line, $\theta\, \alpha/\beta$ (for some new individual constant β) shall appear as a premise on a later line.

4) When a truth-functional inconsistency appears, derive '$P \,\&\, -P$' by rule T, apply ES to transfer dependence to $-\phi$, conditionalize to obtain the theorem $-\phi \rightarrow (P \,\&\, -P)$, and apply T to obtain the theorem ϕ.

It can be shown that if ϕ is valid, then a truth-functional inconsistency will eventually appear in the sequence as above described, but the argument to establish this is too long to be included here.

Instruction 3) requires more careful formulation, but first let us consider as an example theorem 3, page 127. Reducing this valid sentence to prenex normal form we obtain

$$(x)(\exists y)(z)(\exists w)(-Fxy \vee Fwz).$$

To prove this sentence we begin by assuming its negation,

{1}	(1) $-(x)(\exists y)(z)(\exists w)(-Fxy \vee Fwz)$	P

Reducing this to prenex normal form (and omitting the huge number of steps that would be required) we have

{1}	(2) $(\exists x)(y)(\exists z)(w)(Fxy \,\&\, -Fwz)$	

Now, in accord with the two rules under 3), we generate the following sequence:

{3}	(3) $(y)(\exists z)(w)(Fay \,\&\, -Fwz)$	P
{3}	(4) $(\exists z)(w)(Faa \,\&\, -Fwz)$	3 US
{5}	(5) $(w)(Faa \,\&\, -Fwb)$	P
{5}	(6) $Faa \,\&\, -Fab$	5 US
{5}	(7) $Faa \,\&\, -Fbb$	5 US
{3}	(8) $(\exists z)(w)(Fab \,\&\, -Fwz)$	3 US
{9}	(9) $(w)(Fab \,\&\, -Fwc)$	P
{9}	(10) $Fab \,\&\, -Fac$	9 US

Between (6) and (10) we have a truth-functional inconsistency. Continuing, in accord with instruction 4), we complete the derivation as follows:

{5,9}	(11) $P \,\&\, -P$	6,10 T
{3,5}	(12) $P \,\&\, -P$	8,9,11 ES
{3}	(13) $P \,\&\, -P$	4,5,12 ES
{1}	(14) $P \,\&\, -P$	2,3,13 ES

Λ	(15) (1) $\rightarrow$ (P & $-P$)	1,14 C
Λ	(16) $(x)(\exists y)(z)(\exists w)(-Fxy \vee Fwz)$	15 T

Thus, except for the gap between steps (1) and (2), we have a derivation of the sentence on line (16). As mentioned earlier, this gap can be filled by a succession of steps involving rules Q, US, T, UG, and EG (compare example XVII, page 126).

Instruction 3) may be exactly formulated as follows. Beginning with the line on which the prenex normal form of $-\phi$ appears, consider each line in order. (i) If a universal generalization $(\alpha)\theta$ appears on line (n), add (by US) lines on which appear instances $\theta\ \alpha/\beta$ for all individual constants β occurring in sentences on lines up to and including (n), or, if no individual constant occurs in any sentence on lines up to and including (n), add a line on which appears $\theta\ \alpha/\beta$ for some arbitrarily chosen constant β. (ii) If an existential generalization $(\exists\alpha)\theta$ appears on line (n), add (as a premise) a line on which $\theta\ \alpha/\beta$ appears, with a constant β new to the derivation. (iii) If the sentence appearing on line (n) contains an individual constant β not occurring earlier, add (by US) lines on which appear instances $\chi\ \delta/\beta$ of all universal generalizations $(\delta)\chi$ appearing on lines preceding line (n).

If (i) and (iii) both apply to a given line, perform (i) first; similarly, if (ii) and (iii) both apply, perform (ii) first. Thus, in the example both (i) and (iii) apply to line (5); following (i) we added lines (6) and (7), and then in accord with (iii) we added line (8).

Note that in the foregoing example the sequence prescribed by (i)–(iii) would never come to an end (though we did not carry it beyond the point where the inconsistency appeared). In some cases, however, depending upon the order in which universal and existential quantifiers occur in the prefix of the formula from which we begin, the sequence will come to an end after a finite number of steps. For the valid sentence '$(x)((x)Fx \rightarrow Fx)$', a prenex normal form of which is '$(x)(\exists y)(-Fy \vee Fx)$', our generated proof would begin as follows:

{1}	(1) $-(x)(\exists y)(-Fy \vee Fx)$	P
{1}	(2) $(\exists x)(y)(Fy \ \& \ -Fx)$	
{3}	(3) $(y)(Fy \ \& \ -Fa)$	P
{3}	(4) $Fa \ \& \ -Fa$	3 US

Application of (i)–(iii) yields no further lines. For formulas with prefixes that lead to such terminating sequences we therefore have a way of deciding validity: if a truth-functional inconsistency has appeared by the time the sequence has terminated, the original sentence is valid; otherwise it is not. If a sentence ϕ in prenex normal form has a prefix in which no existential quantifier precedes any universal quantifier (in which case a prenex normal form of $-\phi$ may be obtained in which no universal quantifier precedes any existential quantifier), then the sequence will terminate; hence, as noted on page 139, in the case of such sentences we have a decision procedure for validity.

EXERCISES

1. Using metatheorem VIII, construct for each of the following a formula equivalent to its negation (first eliminate ‘ ’ and ‘ ’ if necessary).
   (a) $(x)(Fx \rightarrow Gx)$
   (b) $(\exists x)(Fx \,\&\, Gx)$
   (c) $(\exists x)(Fx \rightarrow Gx)$
   (d) $(x)(\exists y)Fxy$
   (e) $(x)(y)Fxy \leftrightarrow (y)(x)Fxy$
   (f) $(x)Fx \leftrightarrow (\exists y)Gy$
   (g) $(x)(y)(Fxy \rightarrow Fyx)$
   (h) $(x)(y)(z)((Fxy \,\&\, Fyz) \rightarrow Fxz)$
2. Reduce each of the theorems 1–10 (pages 127–8) to prenex normal form.
3. For each of the theorems 1–5, 7–10, 12, 13, 18–21 (pages 127–9), form the dual.
4. Give an example of a theorem that is its own dual.
5. Set forth in detail the argument to demonstrate part 1 of metatheorem IV.
6. Show that Λ is d-consistent if and only if our system of rules is consistent.
7. Given that the system of rules consisting of P, T, C, US, UG, and E is sound, how do we know that the system of rules consisting of P, T, C, US, UG, E, EG, ES, and Q is sound?
8. In the proof of lemma II for the completeness theorem, where did we make use of the assumption that all indices of individual constants occurring in the sentences of Γ are even?
9. By close analogy with Henkin's completeness proof, construct a proof of the following: for any SC sentence ϕ and set of SC sentences Γ, if ϕ is a consequence of Γ, then ϕ is SC derivable from Γ (see Chapter 6).
   (1) Redefine ‘d-consistency’ so that a set of SC sentences Γ is d-consistent iff ‘$P \,\&\, -P$’ is not SC derivable from Γ.
   (2) Redefine ‘maximal d-consistency’ accordingly.
   (3) Show that any d-consistent set of SC sentences Γ is included in a maximal d-consistent set of SC sentences. (This time the list $\phi_1, \phi_2, \phi_3, \ldots, \phi_n, \ldots$ can be any listing whatever of the SC sentences).
   (4) Then show that every maximal d-consistent set of SC sentences is consistent.
   (5) From (3) and (4) deduce the conclusion that, for any set of SC sentences Γ, if Γ is d-consistent, then it is consistent.
   (6) From (5) deduce the metatheorem in question.
10. Find a counter-example to the following assertion: for any set of sentences Γ, if Γ is d-consistent then there is a set of sentences Δ that includes Γ and is maximal d-consistent and ω-complete.

9

IDENTITY AND TERMS

In the present chapter we consider two new artificial languages, $\mathfrak{L}_I$ and $\mathfrak{L}'$, together with their appropriate inference rules. Both are essentially similar to the language $\mathfrak{L}$, but to have enjoyed their advantages earlier would have meant adding further clauses to definitions that are already complicated enough for the beginner. The notation of the language $\mathfrak{L}_I$ is identical with that of $\mathfrak{L}$; by reclassifying the binary predicate 'I^2_1' as a logical constant and interpreting it as standing for the relation of identity, however, we obtain a new language with a new and larger set of logical truths. The language $\mathfrak{L}'$ is obtained from the language $\mathfrak{L}_I$ by adding so-called operation symbols, which serve to denote operations like addition and multiplication and, as will be seen, greatly facilitate the formulation of various familiar theories.

1. *Identity; the language* $\mathfrak{L}_I$. Of all the formulas that logicians have traditionally put forward as laws of logic, surely the most frequently occurring are the following three:

P or not *P*
Not both *P* and not *P*
A is *A*.

Equally traditional, it may be added, are the foolish attacks so often directed against these formulas; the first is alleged to imply that everything is black or white, whereas in fact (as any non-logician would know) some things are gray; the second is supposed to be refuted by the appropriateness of answers like 'Well, it is and it isn't' to questions like 'Is your new job working out satisfactorily?'; and the last, which is most frequently the butt of criticism at this level, is treated as though it asserted that nothing changes.

As suggested earlier, instances of '*P* or not *P*' and 'Not both *P* and not *P*' are usually agreed to be necessary by virtue of their logical form; in other words, all sentences of these forms are necessary. When we come to sentences of the form '*A* is *A*', on the other hand, opinion is divided. According to one view, such sentences are to be regarded necessary by virtue of their logical form; according to another, they are best considered as necessary by virtue of the meaning of the word 'is', in the way in which instances of

> If *A* is warmer than *B* and *B* is warmer than *C*, then *A* is warmer than *C*

are necessary by virtue of the meaning of 'is warmer than'. Clearly the dispute is in large measure terminological; the crux is whether the relation-word 'is' in everyday language should be classified as a logical constant, like 'not', 'or', etc., or whether it should be treated as part of the non-logical vocabulary. In setting up the artificial language $\mathfrak{L}$, we had to make a corresponding choice. We included '$\vee$', '&' and '$-$' among our logical constants but classified all predicates, not excepting 'I^2_1', as non-logical. As a result we obtained as valid sentences

$$P \vee -P$$

and

$$-(P \,\&\, -P),$$

but certainly not

$$I^2_1aa.$$

It is philosophically important for the student to realize clearly to what extent the composition of the set of valid sentences of $\mathfrak{L}$ depends upon how the constants are classified as logical or non-logical, and that this division is rather arbitrary. If we were to reclassify one or more of these symbols (while sticking to the intuitive notion of a valid sentence as one that is true no matter how the non-logical constants are interpreted), we should be led to a quite different group of 'valid' sentences. Suppose, for instance, that we decide to treat the connective '&' as a non-logical constant. This would involve redefining 'interpretation' and 'true' along the following lines: an interpretation would be a non-empty domain $\mathfrak{D}$ together with an assignment that associates with each individual constant of $\mathfrak{L}$ an element of $\mathfrak{D}$, etc. (as on page 56), and that associates with the connective '&' a function from ordered pairs of truth-values to truth-values; 'true under $\mathfrak{I}$' would be defined as on page 60, except that clause 5 would read:

5) if $\phi = (\psi \,\&\, \chi)$ for sentences ψ, χ, then ϕ is true under $\mathfrak{I}$ if and only if the function that $\mathfrak{I}$ assigns to '&' has the value T for the pair of values that $\mathfrak{I}$ gives to ψ and χ.

Still considering a sentence as valid if and only if it is true under all interpretations, we would then find that

$$(P \,\&\, Q) \rightarrow (P \,\&\, Q),$$

for example, was still valid, but that such sentences as

$$(P \,\&\, Q) \rightarrow (Q \,\&\, P)$$

and

$$(P \,\&\, Q) \rightarrow P$$

no longer enjoyed this status. In our regular treatment, '&' in effect has constant denotation as we go from one interpretation to another; it always stands for the truth-function that has the value T for the pair $\langle T, T\rangle$ and the value F for all other pairs; but when we change its status to that of a non-logical constant, we let its denotation vary over the several truth-functions (16 in number) of the appropriate type. The result is a drastic change in the set of valid sentences.

Now if we wish to let our treatment of identity be guided by the intuitive notion that 'is' or 'is the same as' are *not* to count as part of the logical structure of sentences in which they occur, we can develop the theory of identity axiomatically as an independent theory formulated in the artificial language $\mathfrak{L}$. As axioms we take all closures of formulas of the form

$$I^2_1\beta\beta,$$

where β is an individual symbol, and all closures of formulas of the form

$$I^2_1\beta\gamma \rightarrow (\phi \leftrightarrow \psi),$$

where β and γ are individual symbols, ϕ and ψ are atomic formulas, and ϕ is like ψ except that β and γ have been exchanged at one or more places. Intuitively sentences of the former sort will assert that things are identical with themselves; sentences of the latter type represent particular instances of Leibniz's Law: if two things are identical, then whatever is true of the one is true of the other. The assertions or 'theses' of the theory would then be defined as all consequences of these axioms. They would include of course all logical truths of $\mathfrak{L}$, and in addition such sentences as

$$Iaa,$$
$$(x)(y)(Ixy \rightarrow Iyx),$$
$$(x)(y)((Ixy \,\&\, Fx) \rightarrow Fy),$$

which, though not valid, cannot be false as long as 'I' stands for a relation of identity.*

* Some authors restrict the theory of identity to only those theses containing no predicate other than the one representing identity, in our case, 'I^2_1'.

If we approach the matter from the other point of view, we reclassify the predicate 'I_1^2' as a logical constant that always stands for the relation of identity in the domain of the relevant interpretation. (Of course, any other binary predicate could equally well be used). Strictly speaking, in making this small change we shall be constructing another formalized language, which we shall call '$\mathfrak{L}_I$'. The sentences of $\mathfrak{L}_I$ are the same as those of $\mathfrak{L}$, but by appropriate alterations to the definition of 'interpretation' the valid sentences of $\mathfrak{L}_I$ can be made to include the laws of the theory of identity as well as the valid sentences of $\mathfrak{L}$. The resulting logical system, formulated in the language $\mathfrak{L}_I$, is what is usually called the *first order predicate calculus with identity.*

Before going further, it will be useful* to introduce a couple of obvious conventions for writing identity-formulas. Accordingly, let

$\alpha = \beta$	stand for	$I_1^2\alpha\beta$,	and
$\alpha \neq \beta$	stand for	$-I_1^2\alpha\beta$,	

where α,β are any individual symbols of $\mathfrak{L}_I$ (or of $\mathfrak{L}$). Thus, with these conventions the three sentences given above could be written

$$a = a,$$
$$(x)(y)(x = y \rightarrow y = x),$$
$$(x)(y)((x = y \,\&\, Fx) \rightarrow Fy).$$

Defining 'interpretation' for the language $\mathfrak{L}_I$ in such a way as to guarantee that the predicate 'I_1^2' is interpreted as standing for the relation of identity, we say that an *interpretation* of $\mathfrak{L}_I$ consists of a non-empty domain $\mathfrak{D}$ together with an assignment that associates with each individual constant of $\mathfrak{L}_I$ an element of $\mathfrak{D}$, with each n-ary predicate other than 'I_1^2' an n-ary relation among elements of $\mathfrak{D}$, with the predicate 'I_1^2' the identity relation among elements of $\mathfrak{D}$, and with each sentential letter of $\mathfrak{L}_I$ one of the truth-values T or F. (Since 'I_1^2' is now classified as a logical constant, strict parallelism with what we have done before would require deleting the clause 'with the predicate 'I_1^2'', etc., and instead adding a clause to the definition of 'true under $\mathfrak{I}$', guaranteeing that an identity sentence shall be true under $\mathfrak{I}$ if $\mathfrak{I}$ assigns the same individual to both constants appearing therein; but the present method is more economical in certain respects.)

Note that according to this definition every interpretation of $\mathfrak{L}_I$ is an interpretation of $\mathfrak{L}$, but not every interpretation of $\mathfrak{L}$ is an interpretation of $\mathfrak{L}_I$.

The terms 'true', 'valid', 'consequence', and 'consistent' are defined for the language $\mathfrak{L}_I$ in exactly the same way as for $\mathfrak{L}$. It thus turns out that

*It will be useful, but a bit dangerous as well, since the symbol '=' is also employed in the metalanguage; the status of a given occurrence will, it is hoped, always be clear from the context.

every sentence valid in $\mathfrak{L}$ is valid in $\mathfrak{L}_I$, but not vice versa. The sentence '$a = a$' is one of the simplest examples of such a case; it is true under every interpretation of $\mathfrak{L}_I$, but it is not true under every interpretation of $\mathfrak{L}$. Correspondingly, if ϕ is a consequence of Γ relative to $\mathfrak{L}$, then it is also a consequence of Γ relative to $\mathfrak{L}_I$, but the converse does not hold in general. 'Fa' is a consequence of 'Fb' and '$a = b$' relative to $\mathfrak{L}_I$ but not relative to $\mathfrak{L}$. And if a set Γ of sentences is consistent relative to $\mathfrak{L}_I$ it is a consistent set relative to $\mathfrak{L}$, but some sets that are inconsistent relative to $\mathfrak{L}_I$ are consistent relative to $\mathfrak{L}$.

By adding just one rule to our six basic rules of derivation for $\mathfrak{L}$, we obtain an adequate set of rules for $\mathfrak{L}_I$. Proceeding as before, we define a derivation in $\mathfrak{L}_I$ as a finite sequence of consecutively numbered lines, each consisting of a sentence of $\mathfrak{L}_I$ together with a set of numbers (called the premise-numbers of the line), the sequence being constructed according to the rules P, T, C, US, UG, E, and the one additional rule (in which β, γ are individual constants):

I (a) The sentence $\beta = \beta$ may be entered on a line, with the empty set of premise-numbers; and

(b) if a sentence ϕ is like a sentence ψ except that β and γ have been exchanged at one or more places, then ϕ may be entered on a line if ψ and $\beta = \gamma$ appear on earlier lines; as premise-numbers of the new line take those of the earlier lines.

Again as before, a derivation in which a sentence ϕ appears on the last line and all premises of that line belong to a set of sentences Γ is called a *derivation of* ϕ *from* Γ, and a sentence ϕ is *derivable from* a set of sentences Γ if and only if there is a derivation of ϕ from Γ.

The most important property of these concepts is the following: for any sentence ϕ and set of sentences Γ of $\mathfrak{L}_I$: ϕ *is derivable from* Γ *if and only if* ϕ *is a consequence of* Γ. This is the force of the statement that our seven rules are adequate for the first order predicate calculus with identity. The proof is exactly like that given for the predicate calculus without identity, except that in establishing lemma III we specify the interpretation $\mathfrak{I}$ as follows: Divide all the individual constants of $\mathfrak{L}_I$ into classes in such a way that constants α, β are assigned to the same class if and only if the sentence $\alpha = \beta$ is in Δ. By this rule each constant of $\mathfrak{L}_I$ will be assigned to exactly one such class. (We write '$[\gamma]$' for 'the class to which γ belongs', and similarly for other metalinguistic variables.) Let the domain of $\mathfrak{I}$ be the set of all these classes. With each individual constant γ of $\mathfrak{L}_I$ associate as denotation the class $[\gamma]$ of which it is a member. With each sentential letter ϕ of $\mathfrak{L}_I$ associate as denotation the truth-value T if $\phi \in \Delta$, otherwise F. With each n-ary predicate θ of $\mathfrak{L}_I$ associate the set of those n-tuples of classes $\langle[\gamma_1], [\gamma_2], \ldots, [\gamma_n]\rangle$ which are such that the result of writing θ followed by the string $\gamma_1\gamma_2 \ldots \gamma_n$ is an atomic sentence belonging to Δ. (We here assume, what is easily proved, that if $[\gamma_1] = [\delta_1]$, $[\gamma_2] = [\delta_2]$, $\ldots$, $[\gamma_n] = [\delta_n]$, then $\theta\gamma_1\gamma_2 \ldots \gamma_n \in \Delta$ if and only if $\theta\delta_1\delta_2 \ldots \delta_n \in \Delta$.) Note that this interpretation assigns the identity relation to the predicate 'I^2_1'. It is now possible to prove III and thereby the completeness of the predicate calculus with identity. We also get the following version of the Löwenheim-Skolem theorem: if Γ is a consistent set of sentences of $\mathfrak{L}_I$, then Γ is satisfiable by an interpretation having a finite or denumerably infinite domain. (Cp. p. 146).

Finally, with respect to the new language $\mathfrak{L}_I$ we again define a sentence ϕ as a *theorem of logic* if and only if ϕ is derivable from the empty set of sentences. In view of the completeness of our rules it follows that the theorems and the valid sentences coincide.

Here are a few examples of sentences that are theorems of $\mathfrak{L}_I$ but not theorems of $\mathfrak{L}$.

1. $(x)x = x$
2. $(x)(y)(x = y \rightarrow y = x)$

$\{1\}$	(1) $a = b$	P
$\{1\}$	(2) $b = a$	1, I
Λ	(3) $a = b \rightarrow b = a$	1,2 C
Λ	(4) $(y)(a = y \rightarrow y = a)$	3 UG
Λ	(5) $(x)(y)(x = y \rightarrow y = x)$	4 UG

3. $(x)(y)(z)((x = y \,\&\, y = z) \rightarrow x = z)$
4. $(x)(y)((z)(x = z \leftrightarrow y = z) \leftrightarrow x = y)$
5. $(x)(Fx \leftrightarrow (\exists y)(x = y \,\&\, Fy))$
6. $(x)(Fx \leftrightarrow (y)(x = y \rightarrow Fy))$
7. $(x)(y)(x = y \rightarrow (Fx \leftrightarrow Fy))$
8. $(x)(y)((Fx \,\&\, x = y) \leftrightarrow (Fy \,\&\, x = y))$
9. $(\exists x)(y)(Fy \leftrightarrow y = x) \leftrightarrow ((\exists x)Fx \,\&\, (x)(y)((Fx \,\&\, Fy) \rightarrow x = y))$
10. $(x)(\exists y)(y \neq x \,\&\, Fy) \leftrightarrow (\exists x)(\exists y)(x \neq y \,\&\, (Fx \,\&\, Fy))$
11. $((x)(\exists y)Fxy \,\&\, (x) - Fxx) \rightarrow (x)(\exists y)(x \neq y \,\&\, Fxy)$
12. $(Fa \,\&\, -Fb) \rightarrow (\exists x)(\exists y)x \neq y$
13. $(Fa \,\&\, (x)(x \neq a \rightarrow Fx)) \leftrightarrow (x)Fx$
14. $(x)(x \neq a \rightarrow Fx) \rightarrow (x)(y)(x \neq y \rightarrow (Fx \vee Fy))$
15. $(x)(y)(x \neq y \rightarrow (Fx \vee Fy)) \rightarrow ((x)(x \neq a \rightarrow Fx) \vee Fa)$
16. $(\exists x)(y)(y \neq x \rightarrow Fy) \leftrightarrow (x)(y)(x \neq y \rightarrow (Fx \vee Fy))$
17. $(\exists y)(x)x = y \rightarrow ((x)Fx \vee (x) - Fx)$
18. $(x)(y)(z)((x = y \vee x = z) \vee y = z) \rightarrow (((x)Fx \vee (x)(Fx \rightarrow Gx)) \vee (x)(Fx \rightarrow -Gx))$

2. *Parenthetical remarks.* The identity relation for a given domain is a relation that holds only between each element of the domain and itself. Many people find it odd to say that a relation holds between a thing and itself; indeed, this way of talking has a certain resemblance to such non-

sensical questions as 'What's the difference between a duck?', with which children used to amuse themselves. Relations, especially binary relations, have been considered as somehow joining the objects related, and accordingly it has been said that in some sense they are *between* the relata, just as glue is between the boards it holds together. When the matter is conceived in this way, there is of course no room for an identity relation. But there will also be no room for such a relation as that expressed by the words 'is immediately adjacent to', which is the very model of a binary relation. The relation expressed by 'is linked to', applied to the links of a chain, will on the one hand have to be between each link and its next neighbor, and yet reach from one end of the chain to the other. In short, the picture of a binary relation as being literally between its relata is very misleading, to say the least, and is hardly a sound basis upon which to reject the relation of identity.

Reflection on the nature of identity seems to have been part of what led Frege to the sense-denotation distinction. It might be thought, he says, that any sentence of the form

$$A = B,$$

if true, tell us only that something is identical with itself, and, if false, that something is identical with something else. In the former case it would express a triviality; in the latter, an absurdity. Thus every identity-sentence would be either trivial or absurd. But this is impossible to reconcile with the fact that many important scientific discoveries can be stated in just that form. Frege also rejects the view that would explain

$$7 + 5 = 12,$$

for example, as meaning the same as

The expressions '7 + 5' and '12' denote the same object.

He points out in effect that the latter is a contingent truth about language, whereas the former is a necessary truth about numbers. Frege's solution, as has been explained in an earlier chapter, is to distinguish between sense and denotation. The truth of an identity sentence requires only that the two terms have the same denotation; if, in addition, they have different senses the identity sentence will not be trivial, i.e., will not have the same sense as a sentence of the form

$$A = A.$$

Finally it must be pointed out that our own intuitive account of the identity relation is not free of objectionable features. For instance, we have no right to speak of *the* identity relation; by our analysis the identity relation among the elements of one domain will be different from that among the elements of another. Also, we have explicated the term 'relation' in such a way that whatever cannot be a member of a set cannot be related by any relation. Thus insofar as identity is a relation in this sense, such a thing cannot even stand in this relation to itself. This would hold not only of the set of all objects that are not members of themselves, but also of sets described by phrases that give no hint of impending difficulties. The problem is closely related to Russell's Antinomy, and once again every way out seems unintuitive.

3. *Terms; the language* $\mathfrak{L}'$. Although the vocabulary of our artificial languages $\mathfrak{L}$ and $\mathfrak{L}_I$ is sufficiently rich to allow us to formulate many, if not most, of the interesting theories in mathematics and the other sciences, this does not mean that it permits such formulation in the most natural or perspicuous manner. For example, in order to state the simple commutative law for the addition of integers, which usually appears in some such way as the following:

$$(x)(y)\, x + y = y + x,$$

we have to choose a ternary predicate, e.g., 'S^3', interpret it as on page 78, and write

$$(x)(y)(z)(Sxyz \rightarrow Syxz)$$

or perhaps

$$(x)(y)(z)(z_1)((Sxyz \,\&\, Syxz_1) \rightarrow z = z_1).$$

For the more natural formulation we require a new kind of symbol, an *operation symbol* (sometimes called a 'functor' or 'function sign'), to play the role of '+'. The utility of operation symbols becomes even more obvious when one tries to formulate statements involving more than one operation, e.g., the so-called distributive law for multiplication over addition:

$$(x)(y)(z)\, x \cdot (y + z) = x \cdot y + x \cdot z$$

Using the predicates 'S^3' and 'M^3', interpreted as on page 78, we would come out with something like:

$$(x)(y)(z)(w)(w_1)(w_2)(w_3)(w_4)((Syzw \,\&\, Mxww_1 \,\&\, Mxyw_2 \,\&\, Mxzw_3 \,\&\, Sw_2w_3w_4) \rightarrow w_1 = w_4)$$

or

$$(x)(y)(z)(w)((\exists w_1)(Syzw_1 \,\&\, Mxw_1w) \leftrightarrow (\exists w_1)(\exists w_2)(Mxyw_1 \,\&\, Mxzw_2 \,\&\, Sw_1w_2w)).$$

Operation symbols may be incorporated in languages like $\mathfrak{L}$ and $\mathfrak{L}_I$ without any special difficulty, and only a desire to keep unessential complexities to a minimum has prevented us from including them from the start. To make clear just what is involved, we now sketch a language $\mathfrak{L}'$, which is the result of adding operation symbols to $\mathfrak{L}_I$. $\mathfrak{L}'$, together with its appropriate rules of inference, will be called the *first order predicate calculus with identity and operation symbols.*

The *expressions* of the language $\mathfrak{L}'$ are strings (of finite length) of symbols, which in turn are classified as follows:

A. *Variables.* (As on page 44)
B. *Constants.*
  (i) *Logical constants.* (As on page 45, together with the predicate 'I^2_1')
  (ii) *Non-logical constants.*
    (a) *Predicates.* (As on page 45, except that 'I^2_1' is classified as a logical constant)
    (b) *Operation symbols,* which are the lower-case italic letters 'a' through 't', with or without numerical subscripts and superscripts.

An *operation symbol of degree n* (or an *n-ary operation symbol*) is an operation symbol having as superscript a numeral for the positive integer n.

An *individual constant* is an operation symbol without superscript.

The notions *predicate of degree n, sentential letter,* and *individual symbol* are defined as on page 45.

A *term* is an expression that is either an individual symbol or is built up from individual and operation symbols by a finite number of applications of the following rule:

(i) If $\tau_1, \tau_2, \ldots, \tau_n$ are terms and θ is an operation symbol of degree n, then $\theta\tau_1\tau_2 \ldots \tau_n$ (i.e., the result of writing θ, followed by τ_1, followed by $\tau_2, \ldots,$ followed by τ_n) is a term.

An *atomic formula* is an expression that either is a sentential letter or is of the form $\pi\tau_1\tau_2 \ldots \tau_n$, where π is an n-ary predicate and $\tau_1, \tau_2, \ldots, \tau_n$ are terms.

The notions of *formula, sentence,* and *bound* and *free* occurrences of variables in formulas, are defined exactly as on page 45, and the additional syntactic terminology of pages 49 and 50 is carried over unchanged.

We define in addition a *constant term* as a term in which no variable occurs.

Examples. All variables, logical constants, predicates, individual constants, atomic formulas, formulas, and sentences of $\mathfrak{L}_I$ are again variables, logical constants, etc., respectively, of $\mathfrak{L}'$. In addition, the following are terms:

$$a \quad f^1x \quad h^3_2a_1bc_1 \quad g^2xx$$

The following are atomic formulas of $\mathfrak{L}'$ but are not formulas of $\mathfrak{L}_I$:

$$G^2_1f^1ag^1b$$
$$H^1_{16}f^1f^3abf^1c$$

The following are formulas of $\mathfrak{L}'$ but not of $\mathfrak{L}_I$:

$$(x)(y)(P^1f^2xy \to P^1f^2yx)$$
$$(-P \leftrightarrow (x)(G^1f^2xa \to (\exists y)(H^1y \,\&\, (G^1z \vee H^1f_1^2xy))))$$

To give an interpretation of the language $\mathfrak{L}'$ we proceed exactly as we did with respect to $\mathfrak{L}_I$, except that now of course we must assign a denotation to each operation symbol. Accordingly, we say that an *interpretation* of $\mathfrak{L}'$ consists of a non-empty domain $\mathfrak{D}$ together with an assignment that associates

1) with each individual constant of $\mathfrak{L}'$ an element of $\mathfrak{D}$;
2) with each n-ary operation symbol an n-ary operation with respect to $\mathfrak{D}$;
3) with each sentential letter of $\mathfrak{L}'$ one of the truth-values T or F;
4) with each n-ary predicate an n-ary relation among elements of $\mathfrak{D}$; and, in particular,
5) with the binary predicate 'I_1^2', the identity relation among elements of $\mathfrak{D}$.

For problems of translation it will again be useful to have a standard way of giving interpretations. In addition to and by analogy with the notion of English predicate (see page 77) we shall employ a corresponding notion that may be called an 'English descriptor'. An English descriptor is like an English description except that it contains the counter '①', or the counters '①' and '②', or '①', '②', and '③', etc., in one or more places where names or descriptive phrases occur directly. If such an expression is to be used for specifying an operation, it is of course obvious that whenever names of objects of the relevant domain are put in place of the counters the resulting description must denote one and only one element of the domain. (This condition is not sufficient, however.) Thus

① + ②

the only integer different from ①

√①

①² + 2 · ① · ② + ②²

are all English descriptors, and for each we can say that there is at least one interpretation in the standard presentation of which it could play a role. But we also note that for each there is an interpretation with respect to which it does not express an operation. For example, if the domain of the interpretation $\mathfrak{I}$ is the positive integers and '+' has its usual meaning, then

f^2 : ① + ②

would be an appropriate way of saying that $\mathfrak{I}$ associates with 'f^2' the addition operation on positive integers. For, no matter what English expressions denoting positive integers are put in place of '①' and '②' in

① + ②,

the resulting description will name a positive integer. If, however, the domain of $\mathfrak{I}$ consisted of all positive and negative integers except 0, we could not use this English descriptor appropriately to specify an operation, for

$$20 + (-20)$$

would not denote an element of the domain. Contrariwise, if the domain consisted of the set $\{10, 20\}$ we could use the English descriptor

the only integer different from ①

in interpreting a singulary operation symbol, and clearly this depends upon there being exactly two integers in the domain.

Under any interpretation $\mathfrak{I}$ each constant term τ denotes an element of the domain $\mathfrak{D}$ of $\mathfrak{I}$. We define the *value* of a constant term τ under an interpretation $\mathfrak{I}$ as the element it denotes. In other words,

1) if τ is an individual constant, the value of τ under $\mathfrak{I}$ is the element of $\mathfrak{D}$ which $\mathfrak{I}$ assigns to τ;

2) if $\tau = \theta\tau_1 \ldots \tau_n$, where θ is an n-ary operation symbol and $\tau_1, \ldots, \tau_n$ are constant terms, then the value of τ under $\mathfrak{I}$ is the value of the function $\mathfrak{I}(\theta)$—i.e., the function that $\mathfrak{I}$ assigns to θ—when the arguments are the objects that are the values of $\tau_1, \ldots, \tau_n$ under $\mathfrak{I}$.

To obtain a suitable definition of *true under* $\mathfrak{I}$ for the new language $\mathfrak{L}'$ we need make only one change in the definition given for $\mathfrak{L}$ and $\mathfrak{L}_{\text{I}}$ (page 60), namely, to replace its second clause by:

2) if ϕ is atomic and not a sentential letter, then ϕ is true under $\mathfrak{I}$ if and only if the values under $\mathfrak{I}$ of the (constant) terms of ϕ are related (when taken in the order in which their corresponding terms occur in ϕ) by the relation that $\mathfrak{I}$ assigns to the predicate of ϕ.

The definitions of *valid, consequence,* and *consistent* (pages 63–4), as well as those of *normal assignment, tautologous, tautological consequence,* and *truth-functionally consistent* (page 89) are taken over exactly as they stand.

As inference rules for $\mathfrak{L}'$ our old rules P, T, C, US, UG, E, and I will suffice, with minor changes in US and I as follows: in US and I, but *not* in UG, replace all occurrences of 'β' and 'γ' by 'τ' and 'υ', respectively, and understand these latter metalinguistic variables as standing for arbitrary constant terms of $\mathfrak{L}'$.

Derivation, derivation of ϕ *from* Γ, and *derivable* are to be defined as on page 155.

The inference rules as amended are complete: for any sentence ϕ and set of sentences Γ of $\mathfrak{L}'$, ϕ is derivable from Γ if and only if ϕ is a consequence of Γ. In particular, for any sentence ϕ of $\mathfrak{L}'$, ϕ is valid if and only if ϕ is derivable from the empty set.

Examples of the application of the formalized language $\mathfrak{L}'$ and its inference rules will be given in Chapter 11.

EXERCISES

1. Derive theorems 3–18, page 156.
2. (From Kleene, *Introduction to Metamathematics,* p. 408) Construct a derivation of each of the following sentences from the sentence

$$(\exists y)(x)(Fx \leftrightarrow x = y).$$

(a) $(\exists x)(Fx \,\&\, Gx) \leftrightarrow (x)(Fx \rightarrow Gx)$
(b) $(x)Gx \rightarrow (\exists x)(Fx \,\&\, Gx)$
(c) $(\exists y)(Fy \,\&\, Gyy) \leftrightarrow (\exists y)(Fy \,\&\, (\exists x)(Fx \,\&\, Gyx))$
(d) $(\exists x)(Fx \,\&\, (P \rightarrow Gx)) \leftrightarrow (P \rightarrow (\exists x)(Fx \,\&\, Gx))$
(e) $(\exists x)(Fx \,\&\, (Gx \rightarrow P)) \leftrightarrow ((\exists x)(Fx \,\&\, Gx) \rightarrow P)$
(f) $(\exists x)(Fx \,\&\, -Gx) \leftrightarrow -(\exists x)(Fx \,\&\, Gx)$
(g) $(\exists x)(Fx \,\&\, (y)Gxy) \leftrightarrow (y)(\exists x)(Fx \,\&\, Gxy)$

3. For each of the following, either give an interpretation (relative to $\mathfrak{L}_I$) to show that it is not valid, or derive it as a theorem.
(a) $(x)(y)(z)((x \neq y \,\&\, y \neq z) \rightarrow x \neq z)$
(b) $(x)(y)(z)((x \neq y \,\&\, y = z) \rightarrow x \neq z)$
4. From

$$(\exists x)(\exists y)x \neq y \,\&\, (x)(y)(z)((x = y \vee x = z) \vee y = z)$$

derive

$$(\exists x)(\exists y)(x \neq y \,\&\, (z)(z = x \vee z = y)),$$

and conversely.
5. From

$$(\exists x)(x \neq a \,\&\, Fx)$$

derive

$$(\exists x)Fx \,\&\, (Fa \rightarrow (\exists x)(\exists y)(x \neq y \,\&\, (Fx \,\&\, Fy))),$$

and conversely.
6. Using the interpretation given in exercise 1, page 84, with the added entry

I: ① is identical with ②

symbolize the following sentences (use also '=' for '*I*'):
(a) Arthur is a brother of Mary.
(b) Arthur is the only brother of Mary.
(c) The only brother of Mary is the father of Harry.
(d) The only brother of Mary is the father of the only sister of William.
(Hint for (c): Consider 'There is a person who is the father of Harry and who is identical with all those and only those persons who are brothers of Mary.')
7. For distinct variables α, β and formulas ϕ not containing β, let $(\exists_1\alpha)\phi$ be short for $(\exists\beta)(\alpha)(\phi \leftrightarrow \alpha = \beta)$. Thus, e.g., '$(\exists_1 x)Fx$' can be read 'there is exactly one x such that Fx'. Give a formula ϕ containing variables α, β and such that

$$(\exists_1\alpha)(\exists_1\beta)\phi \leftrightarrow (\exists_1\beta)(\exists_1\alpha)\phi$$

is a sentence but is not valid (show the non-validity by giving an interpretation under which the sentence is false).

8. Derive
   (a) '$(x)Fg^1x$' from '$(x)Fx$';
   (b) '$(x)(Hxg \rightarrow Fg^2xg)$' from '$(x)Fx$';
   (c) '$(x)(y)(x = y \rightarrow fx = fy)$' from the empty set;
   (d) '$(x)(y)(z)(w)((x = y \,\&\, z = w) \rightarrow fxz = fyw)$' from the empty set;
   (e) '$(z)(\exists x)(\exists y)z = fxy \rightarrow ((x)(y)Ffxy \rightarrow (x)Fx)$' from the empty set.
9. As plausibly as possible, symbolize in $\mathfrak{L}_I$ the following argument, using the interpretation given.
   $\mathfrak{D}$: the universal set
   C^2: ① can be conceived as greater than ②
   E^1: ① exists
   g: God
   God is that than which nothing greater can be conceived. Try: '$(x)((y) - C^2yx \leftrightarrow x = g)$'). If something does not exist, then something greater than it can be conceived. Therefore, God exists.
   Derive, in $\mathfrak{L}_I$, the symbolized conclusion from the symbolized premises. In what ways is the symbolization unsatisfactory?

10

AXIOMS FOR $\mathfrak{L}_I$

The logical systems presented in Chapters 6, 7, and 9 have been natural deduction systems, consisting of inference rules by means of which one can derive consequences from given assumptions. It is also possible in various ways to systematize the same portions of logic with the help of certain valid sentences taken as logical axioms. In constructing proofs in such systems one is allowed not only to write down sentences that are assumptions or follow from preceding steps by the inference rules, but also at any point to insert one of the logical axioms. The last line of the proof will be a consequence of the assumptions that have been introduced, and, in particular, if the proof contains no assumptions (i.e., if each step is either a logical axiom or follows from predecessors by one of the inference rules), then the last line will be a valid sentence. As will be seen, the use of logical axioms permits a drastic simplification in the rules of inference. On the whole, however, proofs of the most familiar logical laws are somewhat harder to find in the known axiom systems than in natural deduction systems of the type set forth in this book.

As an example of the development of logic from axioms we present in this chapter an axiomatic version of the first order predicate calculus with identity.

1. *Introduction.* The concepts of consequence and validity are semantic concepts, defined in terms of relationships between our formulas and the extralinguistic world. The concepts of derivability and of theorem, on the

other hand, are syntactic. Their definitions refer only to the shapes of the expressions, not to what they may denote when interpreted. As we have seen, for the first order predicate calculus (with or without identity and operation symbols) it is possible to specify a set of performable inference rules which are such that the theorems of logic will coincide with the valid sentences and consequence will coincide with derivability. Thus, though there can be no mechanical way of testing arbitrarily given sentences for validity or consequence, there will be no difficulty in deciding whether any given sequence of sentences is a correct proof or derivation from a given set Γ (provided that Γ is decidable, i.e., that there is a step-by-step procedure for deciding whether or not an arbitrary sentence belongs to Γ). Our rules therefore make it possible for us to establish, in a manner that can be checked by anyone, insight or no insight, that certain sentences are valid.

Another way of characterizing the valid sentences syntactically is by axiomatization. One chooses some easily recognizable valid sentences as axioms, and from these one derives other valid sentences by means of inference rules. Historically this approach has been more common than the one we have followed. Very roughly speaking, the more complex the axioms are, the more simple the rules can be, and vice versa. Our own procedure can be thought of as a limiting case, in which the number of axioms has been reduced to 0 at the price of relative complexity in the content of the rules. To illustrate the opposite situation we shall set forth an axiomatization in which the only inference rules are *modus ponens* and definitional interchange. If we had restricted ourselves to sentences in which the constants '&', 'v', '$\leftrightarrow$' and '$\exists$' do not occur, we could even have eliminated the rule of definitional interchange: every consequence of a set Γ would then be obtainable from Γ and the axioms by a succession of applications of *modus ponens* only.

We must now indicate which sentences are to be taken as logical axioms. For all formulas ϕ, ψ, χ of $\mathfrak{L}_\text{I}$, every variable α, and all individual symbols β, γ, all closures of the following are axioms:

I. $\phi \rightarrow (\psi \rightarrow \phi)$
II. $(\phi \rightarrow (\psi \rightarrow \chi)) \rightarrow ((\phi \rightarrow \psi) \rightarrow (\phi \rightarrow \chi))$
III. $(-\psi \rightarrow -\phi) \rightarrow (\phi \rightarrow \psi)$
IV. $(\alpha)(\phi \rightarrow \psi) \rightarrow ((\alpha)\phi \rightarrow (\alpha)\psi)$
V.* $(\alpha)\phi \rightarrow \phi$
VI. $\phi \rightarrow (\alpha)\phi$, if α does not occur free in ϕ
VII. $(\exists\alpha)\alpha = \beta$
VIII. $\beta = \gamma \rightarrow (\phi \rightarrow \psi)$, where ϕ, ψ are atomic and ψ is like ϕ except for containing an occurrence of γ where ϕ contains an occurrence of β.

* It should be mentioned that, as has been shown by Tarski, the axioms of type V can be proved as theorems on the basis of the remaining axioms; to simplify the deductions, however, I have retained them despite their theoretical redundancy.

Thus there are infinitely many axioms, and indeed infinitely many of each of the eight types mentioned. Perhaps it will be useful to give an example for each type.

Type	*Example*
I.	$(x)(Fxa \rightarrow (Gx \rightarrow Fxa))$
II.	$(x)(y)((P \rightarrow (-Gy \rightarrow Fx)) \rightarrow ((P \rightarrow -Gy) \rightarrow (P \rightarrow Fx)))$
III.	$(-(Fa \rightarrow P) \rightarrow -((x)Fx \rightarrow P)) \rightarrow ((((x)Fx \rightarrow P) \rightarrow (Fa \rightarrow P))$
IV.	$(y)((x)(Fx \rightarrow Gx) \rightarrow ((x)Fx \rightarrow (x)Gx))$
V.	$(x)((x)Fx \rightarrow Fx)$
VI.	$(x)(-Fx \rightarrow (y)-Fx)$
VII.	$(y)(\exists x)x = y$
VIII.	$(x)(x = a \rightarrow (Fxyx \rightarrow Fayx))$

We say that a sentence ϕ follows from sentences ψ and χ by *modus ponens* (MP) (or *detachment*) if and only if $\chi = (\psi \rightarrow \phi)$.

A formula ϕ is *definitionally equivalent* to a formula ψ if and only if there are formulas χ, θ, ϕ_1, ϕ_2, such that ϕ and ψ are alike except that one contains an occurrence of χ at some place where the other contains an occurrence of θ, and either

(1) $\chi = (\phi_1 \vee \phi_2)$ and $\theta = (-\phi_1 \rightarrow \phi_2)$ or
(2) $\chi = (\phi_1 \,\&\, \phi_2)$ and $\theta = -(\phi_1 \rightarrow -\phi_2)$ or
(3) $\chi = (\phi_1 \leftrightarrow \phi_2)$ and $\theta = ((\phi_1 \rightarrow \phi_2) \,\&\, (\phi_2 \rightarrow \phi_1))$ or
(4) $\chi = (\exists\alpha)\phi_1$ and $\theta = -(\alpha)-\phi_1$.

Clearly, if a formula ϕ is definitionally equivalent to a formula ψ, then ψ is definitionally equivalent to ϕ.

A *proof* is a finite sequence of sentences, each of which is an axiom or is definitionally equivalent to an earlier sentence of the sequence or follows from earlier sentences by *modus ponens*.

A sentence ϕ is a *theorem* if and only if ϕ is the last line of a proof. We shall write

$$\vdash \phi$$

as an abbreviation for

all closures of ϕ are theorems,

and similarly for other metalinguistic description-forms.

It will be noted that in this axiomatic system the notion of proof is somewhat simpler than in our natural deduction systems; here a proof is simply a sequence of sentences, whereas in the natural deduction systems it is a sequence of lines, each line consisting of a sentence and a set of premise-numbers.

Instead of giving a list of theorems and proofs, we shall characterize the theorems by means of a series of metatheorems. Unless otherwise specified, the metalinguistic variables 'ϕ', 'ψ', and 'χ' range over formulas, 'α' over variables, 'β' and 'γ' over individual symbols, 'P', 'Q', 'R', and 'T' over expressions. To permit compact statement of such metatheorems as 306 and 307, we shall regard all occurrences of individual constants as 'free' occurrences.

1. If ϕ and $\phi \rightarrow \psi$ are theorems, then ψ is a theorem.

For, given proofs of ϕ and $\phi \rightarrow \psi$, we can construct a proof of ψ by writing the proof of ϕ, followed by that of $\phi \rightarrow \psi$, followed by the single sentence ψ.

2. $\phi \rightarrow \phi$ is a theorem, if ϕ is a sentence.

If ϕ is a sentence, the sequence described as follows will be a proof of $\phi \rightarrow \phi$.

(1)	$\phi \rightarrow ((\phi \rightarrow \phi) \rightarrow \phi)$	I
(2)	$(1) \rightarrow ((\phi \rightarrow (\phi \rightarrow \phi)) \rightarrow (\phi \rightarrow \phi))$	II
(3)	$(\phi \rightarrow (\phi \rightarrow \phi)) \rightarrow (\phi \rightarrow \phi)$	(1)(2)MP
(4)	$\phi \rightarrow (\phi \rightarrow \phi)$	I
(5)	$\phi \rightarrow \phi$	(3)(4)MP

What is given above is of course not itself a proof, but a schematic description of a whole infinity of proofs, one for each sentence ϕ of $\mathfrak{L}_I$. For $\phi =$ 'Fa', as an example, it describes the following proof of '$Fa \rightarrow Fa$'.

(1)	$Fa \rightarrow ((Fa \rightarrow Fa) \rightarrow Fa)$	I
(2)	$(1) \rightarrow ((Fa \rightarrow (Fa \rightarrow Fa)) \rightarrow (Fa \rightarrow Fa))$	II
(3)	$(Fa \rightarrow (Fa \rightarrow Fa)) \rightarrow (Fa \rightarrow Fa)$	(1)(2)MP
(4)	$Fa \rightarrow (Fa \rightarrow Fa)$	I
(5)	$Fa \rightarrow Fa$	(3)(4)MP

The existence of this proof establishes that '$Fa \rightarrow Fa$' is a theorem.

3. $(\psi \rightarrow \chi) \rightarrow ((\phi \rightarrow \psi) \rightarrow (\phi \rightarrow \chi))$ is a theorem, if ϕ, ψ, χ are sentences.

If ϕ, ψ, χ are sentences, the sequence described as follows will be the desired proof.

(1)	$((\phi \rightarrow (\psi \rightarrow \chi)) \rightarrow ((\phi \rightarrow \psi) \rightarrow (\phi \rightarrow \chi))) \rightarrow ((\psi \rightarrow \chi) \rightarrow ((\phi \rightarrow (\psi \rightarrow \chi)) \rightarrow ((\phi \rightarrow \psi) \rightarrow (\phi \rightarrow \chi))))$	I
(2)	$(\phi \rightarrow (\psi \rightarrow \chi)) \rightarrow ((\phi \rightarrow \psi) \rightarrow (\phi \rightarrow \chi))$	II
(3)	$(\psi \rightarrow \chi) \rightarrow ((\phi \rightarrow (\psi \rightarrow \chi)) \rightarrow ((\phi \rightarrow \psi) \rightarrow (\phi \rightarrow \chi)))$	(1)(2)MP
(4)	$(3) \rightarrow (((\psi \rightarrow \chi) \rightarrow (\phi \rightarrow (\psi \rightarrow \chi))) \rightarrow ((\psi \rightarrow \chi) \rightarrow ((\phi \rightarrow \psi) \rightarrow (\phi \rightarrow \chi))))$	II
(5)	$((\psi \rightarrow \chi) \rightarrow (\phi \rightarrow (\psi \rightarrow \chi))) \rightarrow ((\psi \rightarrow \chi) \rightarrow ((\phi \rightarrow \psi) \rightarrow (\phi \rightarrow \chi)))$	(3)(4)MP
(6)	$(\psi \rightarrow \chi) \rightarrow (\phi \rightarrow (\psi \rightarrow \chi))$	I
(7)	$(\psi \rightarrow \chi) \rightarrow ((\phi \rightarrow \psi) \rightarrow (\phi \rightarrow \chi))$	(5)(6)MP

4. If $\phi \rightarrow \psi$ and $\psi \rightarrow \chi$ are theorems, then $\phi \rightarrow \chi$ is a theorem.

Given proofs of $\phi \rightarrow \psi$ and $\psi \rightarrow \chi$, we can construct a proof of $\phi \rightarrow \chi$ by writing the proof of $\phi \rightarrow \psi$, followed by that of $\psi \rightarrow \chi$, followed by that of $(\psi \rightarrow \chi) \rightarrow ((\phi \rightarrow \psi) \rightarrow (\phi \rightarrow \chi))$, followed by the sentence $(\phi \rightarrow \psi) \rightarrow (\phi \rightarrow \chi)$, followed by $\phi \rightarrow \chi$. That this sequence is indeed a proof will be seen by checking the definition of 'proof' and noting that each of the last two lines follows from predecessors by *modus ponens.*

In establishing the next metatheorem and various subsequent metatheorems we shall make explicit use of a type of argument known as *mathematical induction,* of which indeed we have made tacit use earlier. It occurs in several forms. In applying so-called *weak induction* one shows that all the positive integers have some given property P by showing that

(a) 1 has the property P, and
(b) for every positive integer k, if k has the property P, then $k + 1$ has the property P.

Strong induction is also used to show that all the positive integers have some given property P, but in this case one proceeds by establishing that

(a) For each positive integer k, if all positive integers less than k have the property P, then k has the property P.

(By starting with 0 instead of 1, we can of course establish that all natural numbers have the property P). Sometimes weak induction is more convenient; sometimes strong. Essentially the same sort of argument is involved when we show that every positive integer has some property P by arguing that if one of them lacks P, there must be a least such number k, and then by showing that k is neither 1 nor greater than 1. (Compare our argument for lemma III, pp. 145 ff.)

It might seem that by the principle of mathematical induction one could prove only assertions about positive integers and not, as in the case of metatheorem 5, assertions about formulas. To see how the principle applies to these, one must realize that *any* sentence in which a numeral occurs directly, in however subordinate a position, may be regarded as expressing a property of the number denoted by that numeral. Thus

No person under 18 years of age will be issued a permit.

expresses a property of the integer 18, namely, its being a number k such that no person under k years of age will be issued a permit. From a similar point of view, metatheorem 5 may be regarded as an assertion about all positive integers, saying that every positive integer n has the following property: for all formulas ϕ, ψ and expressions Q, $Q(\phi \rightarrow \psi) \rightarrow (Q\phi \rightarrow Q\psi)$ is a theorem if Q is a string of universal quantifiers containing n quantifier-occurrences and containing every variable that occurs free in ϕ or ψ. We shall establish this assertion by using weak induction. Thus, two steps are required. In step (a) we must show that the assertion holds when $n = 1$. In step (b) we show that if it holds when $n = k$, then it holds also when $n = k + 1$. Thereby we prove the assertion for all positive integers n.

5. $Q(\phi \rightarrow \psi) \rightarrow (Q\phi \rightarrow Q\psi)$ is a theorem, if Q is a string of universal quantifiers containing every variable that occurs free in ϕ or ψ.

Let n be the number of quantifier-occurrences in Q.

(a) If $n = 1$, the formula in question is an axiom of type IV.

(b) Suppose that metatheorem 5 holds for $n = k$. Suppose also that Q contains $k + 1$ quantifier occurrences. Then $Q = P(\alpha)$, where P is a string of universal quantifiers containing k quantifier-occurrences, and α is a variable. But

$$P((\alpha)(\phi \rightarrow \psi) \rightarrow ((\alpha)\phi \rightarrow (\alpha)\psi))$$

is an axiom of type IV;

$$P((\alpha)(\phi \rightarrow \psi) \rightarrow ((\alpha)\phi \rightarrow (\alpha)\psi)) \rightarrow (P(\alpha)(\phi \rightarrow \psi) \rightarrow P((\alpha)\phi \rightarrow (\alpha)\psi))$$

is a theorem by hypothesis. Therefore

$$P(\alpha)(\phi \rightarrow \psi) \rightarrow P((\alpha)\phi \rightarrow (\alpha)\psi)$$

is a theorem, by metatheorem 1. But, also by hypothesis,

$$P((\alpha)\phi \rightarrow (\alpha)\psi) \rightarrow (P(\alpha)\phi \rightarrow P(\alpha)\psi)$$

is a theorem, and therefore, by metatheorem 4,

$$P(\alpha)(\phi \rightarrow \psi) \rightarrow (P(\alpha)\phi \rightarrow P(\alpha)\psi)$$

is a theorem.

6. If $Q\phi$ and $Q(\phi \rightarrow \psi)$ are theorems, then $Q\psi$ is a theorem, where Q is a string of universal quantifiers.

This depends upon metatheorem 5 in the way that metatheorem 4 depends upon metatheorem 3.

7. $\vdash (\psi \rightarrow \chi) \rightarrow ((\phi \rightarrow \psi) \rightarrow (\phi \rightarrow \chi))$.

If $(\psi \rightarrow \chi) \rightarrow ((\phi \rightarrow \psi) \rightarrow (\phi \rightarrow \chi))$ is a closure of itself, it is a theorem by metatheorem 3. Any other closure will be of the form $Q((\psi \rightarrow \chi) \rightarrow ((\phi \rightarrow \psi) \rightarrow (\phi \rightarrow \chi)))$. To show that it is a theorem we give an argument analogous to that for metatheorem 3, prefixing 'Q' to each descriptive form and citing metatheorem 6 instead of metatheorem 1.

8. If $Q(\phi \rightarrow \psi)$ and $Q(\psi \rightarrow \chi)$ are theorems, then $Q(\phi \rightarrow \chi)$ is a theorem.

Assume the antecedent of 8. Then $Q((\psi \rightarrow \chi) \rightarrow ((\phi \rightarrow \psi) \rightarrow (\phi \rightarrow \chi)))$ is a sentence and hence, by 7, is a theorem. By two applications of 6, $Q(\phi \rightarrow \chi)$ is a theorem.

9. $\vdash Q(\phi \rightarrow \psi) \rightarrow (Q\phi \rightarrow Q\psi)$.

Let n be the number of quantifier-occurrences in Q.

(a) If $n = 1$, every closure of $Q(\phi \rightarrow \psi) \rightarrow (Q\phi \rightarrow Q\psi)$ is an axiom of type IV and hence a theorem.

(b) Suppose that 9 holds for $n = k$. Suppose that Q contains $k + 1$ quantifier-occurrences. Then $Q = (\alpha)P$, for some variable α and some string P having k quantifier-occurrences. Now if $Q(\phi \to \psi) \to (Q\phi \to Q\psi)$ is a closure of itself, it is a theorem by metatheorem 5. Any other closure will be of the form $T((\alpha)P(\phi \to \psi) \to ((\alpha)P\phi \to (\alpha)P\psi))$. But

$$T((\alpha)(P(\phi \to \psi) \to (P\phi \to P\psi)) \to ((\alpha)P(\phi \to \psi) \to (\alpha)(P\phi \to P\psi)))$$

is an axiom of type IV and hence a theorem. Also,

$$T(\alpha)(P(\phi \to \psi) \to (P\phi \to P\psi))$$

is a theorem, by hypothesis. Hence,

$$T((\alpha)P(\phi \to \psi) \to (\alpha)(P\phi \to P\psi))$$

is a theorem, by metatheorem 6 and the foregoing.

$$T((\alpha)(P\phi \to P\psi) \to ((\alpha)P\phi \to (\alpha)P\psi))$$

is an axiom of type IV and hence a theorem. Therefore,

$$T((\alpha)P(\phi \to \psi) \to ((\alpha)P\phi \to (\alpha)P\psi))$$

is a theorem, by the foregoing and metatheorem 8.

10. $\vdash Q\phi \to \phi$.

Let n be the number of quantifier-occurrences in Q.

(a) If $n = 1$, then every closure of $Q\phi \to \phi$ is an axiom of type V and hence a theorem.

(b) Suppose that 10 holds for $n = k$. Suppose that Q contains $k + 1$ quantifier-occurrences. Then $Q = (\alpha)P$, for some variable α and some string P having k quantifier-occurrences. Let $T((\alpha)P\phi \to \phi)$ be a closure of $Q\phi \to \phi$.

$$T(\alpha)(P\phi \to \phi)$$

is a theorem, by hypothesis.

$$T((\alpha)(P\phi \to \phi) \to ((\alpha)P\phi \to (\alpha)\phi))$$

is a theorem, by 9. Therefore, by the foregoing and 6,

$$T((\alpha)P\phi \to (\alpha)\phi)$$

is a theorem. By hypothesis

$$T((\alpha)\phi \to \phi)$$

is a theorem. Therefore,

$$T((\alpha)P\phi \to \phi)$$

is a theorem, by 8.

If $(\alpha)P\phi \rightarrow \phi$ is a closure of itself, we give an analogous argument, deleting 'T' from the description forms above.

11. $\vdash \phi \rightarrow Q\phi$, if no variable in Q occurs free in ϕ.

The argument is analogous to that for 10, except of course that it relates essentially to axioms of type VI instead of to those of type V.

12. $\vdash Q\phi \rightarrow P\phi$, if every variable in P is either in Q or does not occur free in ϕ.

Assume the antecedent of 12, and let $R(Q\phi \rightarrow P\phi)$ be a closure of $Q\phi \rightarrow P\phi$. Then

$RP(Q\phi \rightarrow \phi)$ is a theorem, by 10;
$R(P(Q\phi \rightarrow \phi) \rightarrow (PQ\phi \rightarrow P\phi))$ is a theorem, by 9; so
$R(PQ\phi \rightarrow P\phi)$ is a theorem, by the foregoing and 6.
$R(Q\phi \rightarrow PQ\phi)$ is a theorem, by 11; and, therefore,
$R(Q\phi \rightarrow P\phi)$ is a theorem, by the foregoing and 8.

If $Q\phi \rightarrow P\phi$ is a closure of itself, we give an analogous argument, deleting 'R' from the description-forms above.

13. If ψ and χ are closures of ϕ, then $\psi \rightarrow \chi$ is a theorem, and χ is a theorem if ψ is a theorem.

Assume that ψ and χ are closures of ϕ. There are four cases to consider: (i) $\psi = \phi$ and $\chi = \phi$; (ii) $\psi = Q\phi$ and $\chi = \phi$; (iii) $\psi = \phi$ and $\chi = P\phi$; (iv) $\psi = Q\phi$ and $\chi = P\phi$. In case (i), $\psi \rightarrow \chi$ is obviously a theorem; in case (ii), it is a theorem by 10; in case (iii), by 11; and in case (iv), by 12. By 1, if ψ is a theorem, χ is a theorem.

It is an obvious corollary of metatheorem 13 that all closures of ϕ are theorems if and only if there is at least one closure of ϕ that is a theorem.

14. If $\vdash \phi$ and $\vdash \phi \rightarrow \psi$, then $\vdash \psi$.

Assume $\vdash \phi$ and $\vdash \phi \rightarrow \psi$. Let $Q\psi$ be a closure of ψ. Let $R\phi$ be a closure of ϕ. Then $RQ\phi$ and $RQ(\phi \rightarrow \psi)$ are theorems; therefore $RQ\psi$ is a theorem, by 6. Therefore $\vdash \psi$, by 13.

15. If $\vdash \phi$ and ψ is definitionally equivalent to ϕ, then $\vdash \psi$.

Assume the antecedent of 15, and let $Q\phi$ be a closure of ϕ. Then $Q\psi$ is definitionally equivalent to $Q\phi$, is a closure of ψ, and is a theorem. Therefore, $\vdash \psi$.

16. If $\vdash \phi \rightarrow \psi$ and $\vdash \psi \rightarrow \chi$, then $\vdash \phi \rightarrow \chi$.

Assume the antecedent of 16. By 7 and two applications of 14, $\vdash \phi \rightarrow \chi$.

17. $\vdash \phi$ if and only if $\vdash Q\phi$. By 10 and 14.

18. If $\vdash \phi \rightarrow \psi$, then $\vdash Q\phi \rightarrow Q\psi$.

For suppose $\vdash \phi \rightarrow \psi$. By 17, $\vdash Q(\phi \rightarrow \psi)$. By 9 and 14, $\vdash Q\phi \rightarrow Q\psi$.

2. *Sentential calculus.* Metatheorems in the 100-series cover, in effect, the sentential calculus; every sentence here declared to be a theorem is either a tautology or a generalization of a tautology.

100. $\vdash \phi \to \phi$

(1) $\vdash \phi \to ((\phi \to \phi) \to \phi)$	I
(2) $\vdash (1) \to ((\phi \to (\phi \to \phi)) \to (\phi \to \phi))$	II
(3) $\vdash (\phi \to (\phi \to \phi)) \to (\phi \to \phi)$	(1)(2)14
(4) $\vdash \phi \to (\phi \to \phi)$	I
(5) $\vdash \phi \to \phi$	(3)(4)14

101. $\vdash -\phi \to (\phi \to \psi)$

(1) $\vdash (-\psi \to -\phi) \to (\phi \to \psi)$	III
(2) $\vdash -\phi \to (-\psi \to -\phi)$	I
(3) $\vdash -\phi \to (\phi \to \psi)$	(1)(2)16

102. $\vdash (\psi \to \chi) \to ((\phi \vee \psi) \to (\phi \vee \chi))$

(1) $\vdash (\psi \to \chi) \to ((-\phi \to \psi) \to (-\phi \to \chi))$	7
(2) $\vdash (\psi \to \chi) \to ((\phi \vee \psi) \to (-\phi \to \chi))$	(1)15
(3) $\vdash (\psi \to \chi) \to ((\phi \vee \psi) \to (\phi \vee \chi))$	(2)15

103. $\vdash --\phi \to \phi$

(1) $\vdash --\phi \to (-\phi \to ---\phi)$	101
(2) $\vdash (-\phi \to ---\phi) \to (--\phi \to \phi)$	III
(3) $\vdash --\phi \to (--\phi \to \phi)$	(1)(2)16
(4) $\vdash (3) \to ((--\phi \to --\phi) \to (--\phi \to \phi))$	II
(5) $\vdash (--\phi \to --\phi) \to (--\phi \to \phi)$	(3)(4)14
(6) $\vdash --\phi \to --\phi$	100
(7) $\vdash --\phi \to \phi$	(5)(6)14

104. $\vdash \phi \to --\phi$

(1) $\vdash ---\phi \to -\phi$	103
(2) $\vdash (---\phi \to -\phi) \to (\phi \to --\phi)$	III
(3) $\vdash \phi \to --\phi$	(1)(2)14

105. $\vdash \phi \to ((\phi \to \psi) \to \psi)$

(1) $\vdash (\phi \to \psi) \to (\phi \to \psi)$	100
(2) $\vdash (1) \to (((\phi \to \psi) \to \phi) \to ((\phi \to \psi) \to \psi))$	II
(3) $\vdash ((\phi \to \psi) \to \phi) \to ((\phi \to \psi) \to \psi)$	(1)(2)14
(4) $\vdash \phi \to ((\phi \to \psi) \to \phi)$	I
(5) $\vdash \phi \to ((\phi \to \psi) \to \psi)$	(3)(4)16

106. $\vdash (\phi \to (\psi \to \chi)) \to (\psi \to (\phi \to \chi))$

(1) $\vdash (\phi \to (\psi \to \chi)) \to ((\phi \to \psi) \to (\phi \to \chi))$	II
(2) $\vdash ((\phi \to \psi) \to (\phi \to \chi)) \to$ $((\psi \to (\phi \to \psi)) \to (\psi \to (\phi \to \chi)))$	7
(3) $\vdash \psi \to (\phi \to \psi)$	I
(4) $\vdash (3) \to (((3) \to (\psi \to (\phi \to \chi))) \to (\psi \to (\phi \to \chi)))$	105
(5) $\vdash ((3) \to (\psi \to (\phi \to \chi))) \to (\psi \to (\phi \to \chi))$	(3)(4)14
(6) $\vdash ((\phi \to \psi) \to (\phi \to \chi)) \to (\psi \to (\phi \to \chi))$	(2)(5)16
(7) $\vdash (\phi \to (\psi \to \chi)) \to (\psi \to (\phi \to \chi))$	(1)(6)16

107. $\vdash (\phi \to \psi) \to ((\psi \to \chi) \to (\phi \to \chi))$ 7,106,14

108. $\vdash \phi \to (\psi \to (\phi \mathbin{\&} \psi))$

(1) $\vdash \phi \to ((\phi \to -\psi) \to -\psi)$	105

(2) $\vdash --(\phi \to -\psi) \to (\phi \to -\psi)$ 103
(3) $\vdash (2) \to (((\phi \to -\psi) \to -\psi) \to$
$(--(\phi \to -\psi) \to -\psi))$ 107
(4) $\vdash ((\phi \to -\psi) \to -\psi) \to (--(\phi \to -\psi) \to -\psi)$ (2)(3)14
(5) $\vdash \phi \to (--(\phi \to -\psi) \to -\psi)$ (1)(4)16
(6) $\vdash (--(\phi \to -\psi) \to -\psi) \to (\psi \to -(\phi \to -\psi))$ III
(7) $\vdash \phi \to (\psi \to -(\phi \to -\psi))$ (5)(6)16
(8) $\vdash \phi \to (\psi \to (\phi \,\&\, \psi))$ (7)15

109. If $\vdash \phi \to \psi$ and $\vdash \psi \to \phi$, then $\vdash \phi \leftrightarrow \psi$. 108,14,15

110. $\vdash (-\phi \to \psi) \to (-\psi \to \phi)$
(1) $\vdash \psi \to --\psi$ 104
(2) $\vdash (1) \to ((-\phi \to \psi) \to (-\phi \to --\psi))$ 7
(3) $\vdash (-\phi \to \psi) \to (-\phi \to --\psi)$ (1)(2)14
(4) $\vdash (-\phi \to --\psi) \to (-\psi \to \phi)$ III
(5) $\vdash (-\phi \to \psi) \to (-\psi \to \phi)$ (3)(4)16

111. $\vdash (-\phi \to \psi) \leftrightarrow (-\psi \to \phi)$ 109,110

112. $\vdash (\phi \to -\psi) \to (\psi \to -\phi)$
(1) $\vdash --\phi \to \phi$ 103
(2) $\vdash (1) \to ((\phi \to -\psi) \to (--\phi \to -\psi))$ 107
(3) $\vdash (\phi \to -\psi) \to (--\phi \to -\psi)$ (1)(2)14
(4) $\vdash (--\phi \to -\psi) \to (\psi \to -\phi)$ III
(5) $\vdash (\phi \to -\psi) \to (\psi \to -\phi)$ (3)(4)16

113. $\vdash (\phi \to -\psi) \leftrightarrow (\psi \to -\phi)$ 112,109

114. $\vdash (\phi \to \psi) \to (-\psi \to -\phi)$ 103,104,107,etc.

115. $\vdash (\phi \to \psi) \leftrightarrow (-\psi \to -\phi)$ 109,114,III

116. $\vdash (\phi \vee \psi) \leftrightarrow (\psi \vee \phi)$ 111,15

117. $\vdash (\phi \,\&\, \psi) \to \phi$ 101,110,14,15

118. $\vdash (\phi \,\&\, \psi) \to \psi$ I,110,14,15

119. $\vdash \phi \leftrightarrow \psi$ if and only if $\vdash \phi \to \psi$ and $\vdash \psi \to \phi$. 109,15,117,118

120. $\vdash \phi \to (\phi \vee \psi)$ 101,104,16,15

121. $\vdash \psi \to (\phi \vee \psi)$ I,15

122. (Replacement) If $\vdash \phi \leftrightarrow \psi$ and χ is like θ except for having an occurrence of ϕ at some place where θ has an occurrence of ψ, then $\vdash \chi \leftrightarrow \theta$ and $\vdash \chi$ if and only if $\vdash \theta$.

Note that if $\vdash \chi \leftrightarrow \theta$, then by 14 and 119, $\vdash \chi$ if and only if $\vdash \theta$.

If $\phi = \chi$, 122 holds trivially. Hence we need only consider the case in which ϕ is a proper subformula of χ.

We demonstrate the metatheorem by mathematical induction (strong) on the order of the formula χ. In other words, we think of 122 as stated in the following form: For every integer n and for all formulas χ, θ, ϕ, ψ, if χ is of order n and ϕ is a proper part of χ and $\vdash \phi \leftrightarrow \psi$ and . . . , then $\vdash \chi \leftrightarrow \theta$.

(a) Assume that the order of χ is 1. Then χ is atomic, and so no formula ϕ is a proper subformula of it. Thus in this case 122 holds trivially.

(b) Assume that 122 holds when the order of χ is equal to or less than k. Assume also the antecedent of 122 and that the order of χ is $k+1$. We have to consider seven cases. Either

(i) $\chi = -\chi_1$, for some formula χ_1 the order of which is k. Then $\theta = -\theta_1$, where χ_1 is like θ_1 except for having an occurrence of ϕ at some place where θ_1 has an occurrence of ψ. By hypothesis or our initial remark in this proof (depending on whether $\phi = \chi_1$ or ϕ is a proper part of χ_1) we have the fact that $\vdash \chi_1 \leftrightarrow \theta_1$. By 115 and 119, $\vdash -\chi_1 \leftrightarrow -\theta_1$, i.e., $\vdash \chi \leftrightarrow \theta$.

(ii) $\chi = (\alpha)\chi_1$, for some variable α and formula χ_1 with order equal to k. Then $\theta = (\alpha)\theta_1$, where χ_1 is like θ_1, etc. By hypothesis or our initial remark, $\vdash \chi_1 \leftrightarrow \theta_1$. By 119 and 18, $\vdash (\alpha)\chi_1 \leftrightarrow (\alpha)\theta_1$, i.e., $\vdash \chi \leftrightarrow \theta$.

(iii) $\chi = (\exists\alpha)\chi_1$. Analogous to case (ii).

(iv) $\chi = \chi_1 \rightarrow \chi_2$, for formulas χ_1, χ_2 with orders equal to or less than k. Then $\theta = \theta_1 \rightarrow \theta_2$, and either $\chi_2 = \theta_2$ and χ_1 is like θ_1, etc., or $\chi_1 = \theta_1$ and χ_2 is like θ_2, etc. In the first case, we have by hypothesis or our initial remark $\vdash \chi_1 \leftrightarrow \theta_1$, and hence by 119 and 107, $\vdash (\chi_1 \rightarrow \chi_2) \leftrightarrow (\theta_1 \rightarrow \theta_2)$. The argument for the second case is similar, except that 7 is used instead of 107. Either way, therefore, $\vdash \chi \leftrightarrow \theta$.

(v) $\chi = \chi_1 \vee \chi_2$. Analogous to case (iv), using 15.

(vi) $\chi = \chi_1 \,\&\, \chi_2$. Analogous to case (v).

(vii) $\chi = \chi_1 \leftrightarrow \chi_2$. Analogous to case (v).

This completes the demonstration of 122.

123. $\vdash \phi \leftrightarrow \phi$ — 100,119

124. $\vdash (\phi \,\&\, \psi) \leftrightarrow (\psi \,\&\, \phi)$

(1) $\vdash -(\phi \rightarrow -\psi) \leftrightarrow -(\phi \rightarrow -\psi)$ — 123

(2) $\vdash -(\phi \rightarrow -\psi) \leftrightarrow -(\psi \rightarrow -\phi)$ — (1),122,113

(3) $\vdash -(\phi \rightarrow -\psi) \leftrightarrow (\psi \,\&\, \phi)$ — (2),15

(4) $\vdash (\phi \,\&\, \psi) \leftrightarrow (\psi \,\&\, \phi)$ — (3),15

125. $\vdash (\phi \leftrightarrow \psi) \leftrightarrow (\psi \leftrightarrow \phi)$ — 123,15,124,122

126. $\vdash (\phi \leftrightarrow \psi) \leftrightarrow (-\phi \leftrightarrow -\psi)$ — 125,15,122,115

127. $\vdash \phi \leftrightarrow --\phi$ — 109,103,104

128. $\vdash (\phi \rightarrow (\psi \rightarrow \chi)) \leftrightarrow (\psi \rightarrow (\phi \rightarrow \chi))$ — 106,119

129. $\vdash (-\phi \rightarrow \phi) \rightarrow \phi$

(1) $\vdash -\phi \rightarrow (\phi \rightarrow -(\phi \rightarrow \phi))$ — 101

(2) $\vdash (1) \rightarrow ((-\phi \rightarrow \phi) \rightarrow (-\phi \rightarrow -(\phi \rightarrow \phi)))$ — II

(3) $\vdash (-\phi \rightarrow \phi) \rightarrow (-\phi \rightarrow -(\phi \rightarrow \phi))$ — (1)(2)14

(4) $\vdash (-\phi \rightarrow -(\phi \rightarrow \phi)) \rightarrow ((\phi \rightarrow \phi) \rightarrow \phi)$ — III

(5) $\vdash (\phi \rightarrow \phi) \rightarrow ((-\phi \rightarrow -(\phi \rightarrow \phi)) \rightarrow \phi)$ — (4)128,122

(6) $\vdash \phi \rightarrow \phi$ — 100

(7) $\vdash (-\phi \rightarrow -(\phi \rightarrow \phi)) \rightarrow \phi$ — (5)(6)14

(8) $\vdash (-\phi \rightarrow \phi) \rightarrow \phi$ — (3)(7)

130. $\vdash \phi \leftrightarrow (\phi \vee \phi)$ — 120,129,15,119

131. $\vdash \phi \leftrightarrow (\phi \,\&\, \phi)$
 (1) $\vdash -\phi \leftrightarrow (-\phi \vee -\phi)$ 130
 (2) $\vdash --\phi \leftrightarrow -(-\phi \vee -\phi)$ (1)122,126
 (3) $\vdash --\phi \leftrightarrow -(--\phi \rightarrow -\phi)$ (2)15
 (4) $\vdash \phi \leftrightarrow -(\phi \rightarrow -\phi)$ (3)127,122
 (5) $\vdash \phi \leftrightarrow (\phi \,\&\, \phi)$ (4)

132. $\vdash -(\phi \,\&\, \psi) \leftrightarrow (-\phi \vee -\psi)$ 122,15,127

133. $\vdash -(\phi \vee \psi) \leftrightarrow (-\phi \,\&\, -\psi)$

134. $\vdash (\phi \,\&\, \psi) \leftrightarrow -(-\phi \vee -\psi)$

135. $\vdash (\phi \vee \psi) \leftrightarrow -(-\phi \,\&\, -\psi)$

136. $\vdash (\phi \vee (\psi \vee \chi)) \leftrightarrow ((\phi \vee \psi) \vee \chi)$ 128,15,122,116

137. $\vdash (\phi \,\&\, (\psi \,\&\, \chi)) \leftrightarrow ((\phi \,\&\, \psi) \,\&\, \chi)$

138. $\vdash (\phi \rightarrow (\psi \rightarrow \chi)) \leftrightarrow ((\phi \,\&\, \psi) \rightarrow \chi)$ 136,122,132,15

139. $\vdash (\phi \rightarrow \psi) \rightarrow ((\phi \rightarrow \chi) \rightarrow (\phi \rightarrow (\psi \,\&\, \chi)))$ 108,7,14,16

140. $\vdash (\phi \vee (\psi \,\&\, \chi)) \leftrightarrow ((\phi \vee \psi) \,\&\, (\phi \vee \chi))$
 (1) $\vdash (\psi \,\&\, \chi) \rightarrow \psi$ 117
 (2) $\vdash (1) \rightarrow ((\phi \vee (\psi \,\&\, \chi)) \rightarrow (\phi \vee \psi))$ 102
 (3) $\vdash (\phi \vee (\psi \,\&\, \chi)) \rightarrow (\phi \vee \psi)$ (1)(2)14
 (4) $\vdash (\psi \,\&\, \chi) \rightarrow \psi$ 117
 (5) $\vdash (4) \rightarrow ((\phi \vee (\psi \,\&\, \chi)) \rightarrow (\phi \vee \psi))$ 102
 (6) $\vdash (\phi \vee (\psi \,\&\, \chi)) \rightarrow (\phi \vee \psi)$ (4)(5)14
 (7) $\vdash (3) \rightarrow ((6) \rightarrow ((\phi \vee (\psi \,\&\, \chi)) \rightarrow ((\phi \vee \psi) \,\&\, (\phi \vee \chi))))$ 139
 (8) $\vdash (\phi \vee (\psi \,\&\, \chi)) \rightarrow ((\phi \vee \psi) \,\&\, (\phi \vee \chi))$ (3)(6)(7)14
 (9) $\vdash \psi \rightarrow (\chi \rightarrow (\psi \,\&\, \chi))$ 108
 (10) $\vdash (9) \rightarrow ((-\phi \rightarrow \psi) \rightarrow (-\phi \rightarrow (\chi \rightarrow (\psi \,\&\, \chi))))$ 7
 (11) $\vdash (-\phi \rightarrow \psi) \rightarrow (-\phi \rightarrow (\chi \rightarrow (\psi \,\&\, \chi)))$ (9)(10)14
 (12) $\vdash (-\phi \rightarrow (\chi \rightarrow (\psi \,\&\, \chi))) \rightarrow ((-\phi \rightarrow \chi) \rightarrow$
 $(-\phi \rightarrow (\psi \,\&\, \chi)))$ 11
 (13) $\vdash (-\phi \rightarrow \psi) \rightarrow ((-\phi \rightarrow \chi) \rightarrow (-\phi \rightarrow (\psi \,\&\, \chi)))$ (11)(12)16
 (14) $\vdash (\phi \vee \psi) \rightarrow ((\phi \vee \chi) \rightarrow (\phi \vee (\psi \,\&\, \chi)))$ (13)15
 (15) $\vdash ((\phi \vee \psi) \,\&\, (\phi \vee \chi)) \rightarrow (\phi \vee (\psi \,\&\, \chi))$ (14)122,138
 (16) $\vdash (\phi \vee (\psi \,\&\, \chi)) \leftrightarrow ((\phi \vee \psi) \,\&\, (\phi \vee \chi))$ (8)(15)119

141. $\vdash (\phi \,\&\, (\psi \vee \chi)) \leftrightarrow ((\phi \,\&\, \psi) \vee (\phi \,\&\, \chi))$ 140,122,126, 133–5

142. $\vdash (\phi \rightarrow \psi) \rightarrow ((\chi \rightarrow \theta) \rightarrow ((\phi \,\&\, \chi) \rightarrow (\psi \,\&\, \theta)))$

143. $\vdash (\phi \rightarrow \psi) \rightarrow ((\chi \rightarrow \psi) \rightarrow ((\phi \vee \chi) \rightarrow \psi))$

144. $\vdash (\phi \rightarrow (\phi \rightarrow \psi)) \leftrightarrow (\phi \rightarrow \psi)$

145. $\vdash (\phi \rightarrow (\psi \leftrightarrow \chi)) \leftrightarrow ((\phi \rightarrow (\psi \rightarrow \chi)) \,\&\, (\phi \rightarrow (\chi \rightarrow \psi)))$

146. $\vdash (\phi \,\&\, -\phi) \rightarrow \psi$

3. *Quantification theory*. In this section we shall use the variables 'P' and 'Q' for arbitrary strings of universal quantifiers, and 'P^e' and 'Q^e' for the

corresponding strings of existential quantifiers. Thus, if $Q =$ '$(x)(y)(z)$', $Q^e =$ '$(\exists x)(\exists y)(\exists z)$'.

200. $\vdash Q-\phi \rightarrow -Q\phi$

(1) $\vdash Q\phi \rightarrow \phi$	10
(2) $\vdash -\phi \rightarrow -Q\phi$	(1)122,115
(3) $\vdash Q-\phi \rightarrow Q-Q\phi$	(2)18
(4) $\vdash Q-Q\phi \rightarrow -Q\phi$	10
(5) $\vdash Q-\phi \rightarrow -Q\phi$	(3)(4)16

201. $\vdash(\phi \rightarrow Q\psi) \rightarrow Q(\phi \rightarrow \psi)$, if no variable in Q is free in ϕ. Assume the antecedent; then

(1) $\vdash Q\psi \rightarrow \psi$	10
(2) $\vdash(1) \rightarrow ((\phi \rightarrow Q\psi) \rightarrow (\phi \rightarrow \psi))$	7
(3) $\vdash(\phi \rightarrow Q\psi) \rightarrow (\phi \rightarrow \psi)$	(1)(2)14
(4) $\vdash Q(\phi \rightarrow Q\psi) \rightarrow Q(\phi \rightarrow \psi)$	(3)18
(5) $\vdash(\phi \rightarrow Q\psi) \rightarrow Q(\phi \rightarrow Q\psi)$	11
(6) $\vdash(\phi \rightarrow Q\psi) \rightarrow Q(\phi \rightarrow \psi)$	(4)(5)16

202. $\vdash Q(\phi \rightarrow \psi) \rightarrow (\phi \rightarrow Q\psi)$, if no variable in Q is free in ϕ. Assume the antecedent; then

(1) $\vdash Q(\phi \rightarrow \psi) \rightarrow (Q\phi \rightarrow Q\psi)$	9
(2) $\vdash \phi \rightarrow Q\phi$	11
(3) $\vdash Q\phi \rightarrow (Q(\phi \rightarrow \psi) \rightarrow Q\psi)$	(1)122,128
(4) $\vdash \phi \rightarrow (Q(\phi \rightarrow \psi) \rightarrow Q\psi)$	(2)(3)16
(5) $\vdash Q(\phi \rightarrow \psi) \rightarrow (\phi \rightarrow Q\psi)$	(4)122,128

203. $\vdash \phi \leftrightarrow Q\phi$, if no variable in Q is free in ϕ. — 10,11

204. $\vdash Q\phi \leftrightarrow -Q^e-\phi$

Let n be the number of quantifier-occurrences in Q.

(a) If $n = 1$, $Q = (\alpha)$ and $Q^e = (\exists\alpha)$, for some variable α.

(1) $\vdash(\alpha)\phi \leftrightarrow (\alpha)\phi$	123
(2) $\vdash(\alpha)\phi \leftrightarrow --(\alpha)--\phi$	(1)122,127
(3) $\vdash(\alpha)\phi \leftrightarrow -(\exists\alpha)-\phi$	(2)15

(b) Assume 204 for $n = k$. Suppose that Q contains $k + 1$ quantifier-occurrences. Then $Q = P(\alpha)$ and $Q^e = P^e(\exists\alpha)$, for some variable α and string P of k quantifier-occurrences.

(1) $\vdash P(\alpha)\phi \leftrightarrow P(\alpha)\phi$	123
(2) $\vdash P(\alpha)\phi \leftrightarrow -P^e-(\alpha)\phi$	(1)122,hp.
(3) $\vdash P(\alpha)\phi \leftrightarrow -P^e-(\alpha)--\phi$	(2)122,127
(4) $\vdash P(\alpha)\phi \leftrightarrow -P^e(\exists\alpha)-\phi$	(3)15

205. $\vdash -Q-\phi \leftrightarrow Q^e\phi$	204,122,123
206. $\vdash -Q\phi \leftrightarrow Q^e-\phi$	204,122,123
207. $\vdash Q-\phi \leftrightarrow -Q^e\phi$	204,122,123
208. $\vdash \phi \rightarrow Q^e\phi$	10,122,113,205
209. $\vdash P\phi \rightarrow Q^e\phi$	208,10,16
210. If $\vdash \phi$, then $\vdash Q^e\phi$	17,209,16

211. If $\vdash \phi \rightarrow \psi$, then $\vdash Q^e\phi \rightarrow Q^e\psi$ — 122,115,18,115 205

212. $\vdash P^e\phi \rightarrow Q^e\phi$, if every variable in P is either in Q or does not occur in ϕ. — 12,122,115,205

213. $\vdash Q^e\phi \rightarrow \phi$, if no variable in Q is free in ϕ. — 11,122,113,205

214. If $\vdash \phi \rightarrow \psi$ and no variable in Q is free in ϕ, then $\vdash \phi \rightarrow Q\psi$. — 18,11,16

215. $\vdash P^eQ\phi \rightarrow QP^e\phi$

(1)	$\vdash \phi \rightarrow P^e\phi$	208
(2)	$\vdash Q\phi \rightarrow QP^e\phi$	(1)18
(3)	$\vdash P^eQ\phi \rightarrow P^eQP^e\phi$	(2)211
(4)	$\vdash P^eQP^e\phi \rightarrow QP^e\phi$	213
(5)	$\vdash P^eQ\phi \rightarrow QP^e\phi$	(3)(4)16

216. $\vdash Q(\phi \,\&\, \psi) \leftrightarrow (Q\phi \,\&\, Q\psi)$ — 117,118,18,138, 108,etc.

217. $\vdash Q(\phi \leftrightarrow \psi) \rightarrow (Q\phi \leftrightarrow Q\psi)$

218. $\vdash (Q\phi \vee Q\psi) \rightarrow Q(\phi \vee \psi)$

219. $\vdash Q(\phi \rightarrow \psi) \rightarrow (Q^e\phi \rightarrow Q^e\psi)$

220. $\vdash Q^e(\phi \vee \psi) \leftrightarrow (Q^e\phi \vee Q^e\psi)$

221. $\vdash Q^e(\phi \,\&\, \psi) \rightarrow (Q^e\phi \,\&\, Q^e\psi)$

222. $\vdash Q^e(\phi \rightarrow \psi) \leftrightarrow (Q\phi \rightarrow Q^e\psi)$

223. $\vdash Q(\phi \vee \psi) \rightarrow (Q^e\phi \vee Q\psi)$

224. $\vdash (P\phi \vee Q^e\psi) \rightarrow Q^e(\phi \vee \psi)$

225. $\vdash ((Q^e\phi \rightarrow Q\psi) \rightarrow Q(\phi \rightarrow \psi))$

226. $\vdash ((Q^e\phi \rightarrow Q^e\psi) \rightarrow Q^e(\phi \rightarrow \psi))$

227. $\vdash (Q\phi \rightarrow Q\psi) \rightarrow Q^e(\phi \rightarrow \psi)$

228. $\vdash (Q\phi \,\&\, Q^e\psi) \rightarrow Q^e(\phi \,\&\, \psi)$

229. $\vdash Q^e\phi \leftrightarrow \phi$, if no variable in Q is free in ϕ.

230. $\vdash Q(\phi \,\&\, \psi) \leftrightarrow (\phi \,\&\, Q\psi)$, if, etc.

231. $\vdash Q^e(\phi \,\&\, \psi) \leftrightarrow (\phi \,\&\, Q^e\psi)$, if, etc.

232. $\vdash Q(\phi \vee \psi) \leftrightarrow (\phi \vee Q\psi)$, if, etc.

233. $\vdash Q^e(\phi \vee \psi) \leftrightarrow (\phi \vee Q^e\psi)$, if, etc.

234. $\vdash Q(\psi \rightarrow \phi) \leftrightarrow (Q^e\psi \rightarrow \phi)$, if, etc.

235. $\vdash Q^e(\psi \rightarrow \phi) \leftrightarrow (Q\psi \rightarrow \phi)$, if, etc.

236. $\vdash Q^e(\phi \rightarrow Q\phi)$

4. *Identity, further quantification theory, and substitution.*

300. $\vdash \alpha = \beta \rightarrow (\alpha = \gamma \rightarrow \beta = \gamma)$ — VIII

301. $\vdash \alpha = \alpha$

(1)	$\vdash \beta = \alpha \rightarrow (\beta = \alpha \rightarrow \alpha = \alpha)$	300
(2)	$\vdash \beta = \alpha \rightarrow \alpha = \alpha$	(1)122,144
(3)	$\vdash (\beta)(\beta = \alpha \rightarrow \alpha = \alpha)$	(2)17

(4) $\vdash(\exists\beta)\beta = \alpha \rightarrow \alpha = \alpha$	(3)122,234
(5) $\vdash(\exists\beta)\beta = \alpha$	VII
(6) $\vdash \alpha = \alpha$	(4)(5)14

302. $\vdash \alpha = \beta \rightarrow \beta = \alpha$

(1) $\vdash \alpha = \beta \rightarrow (\alpha = \alpha \rightarrow \beta = \alpha)$	300
(2) $\vdash \alpha = \alpha \rightarrow (\alpha = \beta \rightarrow \beta = \alpha)$	(1)122,128
(3) $\vdash \alpha = \alpha$	301
(4) $\vdash \alpha = \beta \rightarrow \beta = \alpha$	(2)(3)14

303. $\vdash \alpha = \beta \rightarrow (\beta = \gamma \rightarrow \alpha = \gamma)$ 300,302,16

304. $\vdash \beta = \gamma \rightarrow (\phi \leftrightarrow \psi)$, where ϕ,ψ are atomic and ψ is like ϕ except for containing occurrences of γ at one or more places where ϕ contains occurrences of β. From VIII, by induction on the number of places where ϕ has β and ψ has γ.

305. $\vdash \beta = \gamma \rightarrow (\phi \leftrightarrow \psi)$, where ψ is like ϕ except for containing free occurrences of γ at one or more places where ϕ contains free occurrences of β. From 304, by induction on the order of ϕ.

306. $\vdash(\alpha)\phi \rightarrow \psi$, if ψ is like ϕ except for containing free occurrences of β wherever ϕ contains free occurrences of α.

If $\alpha = \beta$, 306 holds by 10. If $\alpha \neq \beta$, we argue as follows, using 'Hp306' as short for 'hypothesis of 306'.

(1) $\vdash \alpha = \beta \rightarrow (\phi \rightarrow \psi)$	HP306,305,145,etc.
(2) $\vdash \phi \rightarrow (\alpha = \beta \rightarrow \psi)$	(1),106,119,122
(3) $\vdash(\alpha)\phi \rightarrow (\alpha)(\alpha = \beta \rightarrow \psi)$	(2),18
(4) $\vdash(\alpha)\phi \rightarrow ((\exists\alpha)\alpha = \beta \rightarrow \psi)$	(3),122,234
(5) $\vdash(\exists\alpha)\alpha = \beta \rightarrow ((\alpha)\phi \rightarrow \psi)$	(4),106,119,122
(6) $\vdash(\exists\alpha)\alpha = \beta$	VII
(7) $\vdash(\alpha)\phi \rightarrow \psi$	(5)(6)14

307. $\vdash Q\phi \rightarrow \psi$ and $\vdash \psi \rightarrow Q^e\phi$, if ψ is like ϕ except for containing free occurrences of $\beta_1, \ldots, \beta_n$ (not necessarily distinct) where ϕ contains, respectively, the distinct variables $\alpha_1, \ldots, \alpha_n$, and $\alpha_1, \ldots, \alpha_n$ all occur in Q. From 306, by induction on the number of quantifier-occurrences.

308. $\vdash(\alpha)\phi \leftrightarrow (\beta)\psi$, if $(\alpha)\phi$ and $(\beta)\psi$ are alike except that the former has occurrences of the variable α where and only where the latter has occurrences of the variable β.

309. (Substitution) If ϕ is a theorem and ψ is a substitution-instance of ϕ, then ψ is a theorem.

To establish this metatheorem we must first, of course, define the crucial term 'substitution-instance'. An auxiliary notion is needed, that of a stencil. A *stencil* is an expression that is either a sentence of $\mathfrak{L}_{\mathrm{I}}$ or is obtainable from such a sentence by putting the counters '①', or '①' and '②', or '①', '②', and '③', etc., where that sentence contains occurrences of individual constants. Thus the following are stencils:

$$Fa①②$$
$$F①②①$$
$$(x)(Fx①a \rightarrow (\exists y)Gy②b);$$

but

$$Fx①②$$

and

$$Fa①③$$

are not. If a stencil contains no occurrence of '①' (i.e., if it is a sentence), it is of *degree 0*; if it contains an occurrence of '①' but no occurrence of '②', it is of *degree 1*; if it contains an occurrence of '②' but no occurrence of '③', it is of *degree 2*, and so on. If σ is a stencil of degree $n(n > 0)$ and $\delta_1, \ldots, \delta_n$ are individual symbols, then $\sigma(\delta_1, \ldots, \delta_n)$ is the result of putting δ_1 for '①', δ_2 for '②', δ_3 for '③', and so on, in σ.

Now, let ϕ be a sentence of $\mathfrak{L}_I$, and let θ be an n-ary predicate (other than the identity predicate 'I^2_1') that occurs in ϕ. Further, let σ be a stencil of degree n. To *substitute* σ for θ in ϕ we proceed as follows: for each occurrence of $\theta\delta_1 \ldots \delta_n$ in ϕ, where $\delta_1, \ldots, \delta_n$ are individual symbols, we put $\sigma(\delta_1, \ldots, \delta_n)$. If in this substitution we have at all places obeyed the rule that *no variable shall occur bound at any place in any substituted* $\sigma(\delta_1, \ldots, \delta_n)$ *unless it already occurs bound at that place in* σ,* then we say that the substitution is *legitimate*. (If θ is a sentential letter and σ is a stencil of degree 0, then to substitute σ for θ in ϕ we simply replace each occurrence of θ in ϕ by an occurrence of σ; all such substitutions are also to be considered legitimate).

A sentence ψ is a *substitution-instance* of a sentence ϕ if and only if ψ can be obtained from ϕ by a legitimate substitution. Metatheorem 309 thus says that whatever can be obtained from a theorem ϕ by legitimate substitution for a predicate θ occurring therein, is itself a theorem.

Example 1. Substitute '$G①$ & $Ha①$' for 'F' in '$(x)(\exists y)(Fx \rightarrow (Gx \& Fy))$'. For '$Fx$' we put '$Gx \& Hax$'; for '$Fy$' we put '$Gy \& Hay$'; thus obtaining the sentence

$$(x)(\exists y)((Gx \& Hax) \rightarrow (Gx \& (Gy \& Hay))).$$

The substitution is legitimate.

Example 2. Substitute '$(\exists y)G①y$' for 'F' in '$(y)((x)Fx \rightarrow Fy)$'. For '$Fx$' we put '$(\exists y)Gxy$' and for '$Fy$' we put '$(\exists y)Gyy$', obtaining

$$(y)((x)(\exists y)Gxy \rightarrow (\exists y)Gyy).$$

The substitution is not legitimate, because for 'Fy' ($= \theta\delta_1$) we put '$(\exists y)Gyy$' ($= \sigma(\delta_1)$), in which 'y' occurs bound at the sixth place (counting signs

*Since the stencil σ is not ordinarily a formula of $\mathfrak{L}_I$, we should more properly say 'unless it already occurs bound at that place in a sentence from which the stencil σ is derived'.

from the left) although it does not occur bound at the sixth place in any sentence from which the stencil '$(\exists y)G①y$' may be obtained. Note that, although the sentence '$(y)((x)Fx \rightarrow Fy)$' is valid, the result of the substitution is not valid.

Example 3. Substitute '$G①② \rightarrow G②①$' for 'G' in '$(x)(y)(Gxy \rightarrow Gyx)$'. For '$Gxy$' we put '$Gxy \rightarrow Gyx$', and for '$Gyx$' we put '$Gyx \rightarrow Gxy$', obtaining as a result of the substitution the sentence

$$(x)(y)((Gxy \rightarrow Gyx) \rightarrow (Gyx \rightarrow Gxy)).$$

Again the substitution is legitimate.

The argument for the correctness of the metatheorem is in outline as follows. All substitution-instances of axioms of type I are again axioms of type I; similarly for axioms of types II–VI. There are no substitution-instances of axioms of type VII, since these contain no non-logical predicates. All substitution-instances of axioms of type VIII are theorems, as is evident from metatheorem 305. Further, the result of making a single substitution throughout a *modus ponens* argument is again a *modus ponens* argument, and the same is true for a definitional interchange. Now suppose that ϕ is a theorem, and that ψ results from ϕ by making a legitimate substitution of a stencil σ for a predicate θ in ϕ. Consider the proof of ϕ. If σ can be substituted legitimately for θ throughout this proof, we can thereby get a proof of ψ, for axioms go into axioms (except that axioms of type VIII go into theorems, which may then be replaced by their proofs), applications of *modus ponens* go into applications of *modus ponens,* and applications of definitional interchange go into applications of definitional interchange. If σ cannot be substituted legitimately for θ in all steps of the proof of ϕ, we have to take a detour on the way to our destination. First we obtain another stencil σ', which is like σ except that all bound variables have been rewritten with distinct bound variables in such a way that no bound variable in σ' occurs in any step of the proof of ϕ. Then, throughout the proof of ϕ we substitute σ' for θ (and these substitutions will be legitimate), obtaining, as before, a proof of a substitution instance ψ' of ϕ. Since ψ' is a theorem, it follows by 308 and 122 that ψ is a theorem.

5. *Proofs from assumptions.* It is natural to enlarge our notion of proof in the following way. We shall say that a finite sequence of sentences is a *proof of the sentence* ϕ *from the assumptions* Γ (where Γ is a set of sentences) if and only if ϕ is the last term of the sequence and each term of the sequence is either an axiom or one of the assumptions Γ or follows from earlier terms by *modus ponens* or definitional interchange. The notation

$$\Gamma \Vdash \phi$$

will be short for

there is a proof of ϕ from the assumptions Γ,

and similarly for other description forms representing sentences and sets of sentences.

From the definition above we have as immediate consequences, where ϕ and ψ are sentences, and Γ and Δ are sets of sentences:

400. $\Lambda \Vdash \phi$ if and only if ϕ is a theorem.

401. If $\phi \in \Gamma$, then $\Gamma \Vdash \phi$.

402. If $\Gamma \Vdash \phi$ and $\Gamma \subset \Delta$, then $\Delta \Vdash \phi$.

403. If $\Gamma \Vdash \phi$ and $\Delta \Vdash \phi \rightarrow \psi$, then $\Gamma \cup \Delta \Vdash \psi$.

404. If ϕ is definitionally equivalent to ψ, then $\Gamma \Vdash \phi$ if and only if $\Gamma \Vdash \psi$.

Not quite so obvious is the following important metatheorem, the so-called Deduction Theorem.

405. $\Gamma \cup \{\phi\} \Vdash \psi$ if and only if $\Gamma \Vdash \phi \rightarrow \psi$.

Assume $\Gamma \Vdash \phi \rightarrow \psi$. By 402, $\Gamma \cup \{\phi\} \Vdash \phi \rightarrow \psi$. By 403, $\Gamma \cup \{\phi\} \Vdash \psi$.

Now assume $\Gamma \cup \{\phi\} \Vdash \psi$. This means that there is a finite sequence θ_1, $\theta_2, \ldots, \theta_n = \psi$ that is a proof of ψ from the assumptions $\Gamma \cup \{\phi\}$. Thus, for each i from 1 to n, θ_i is either ϕ or it is an axiom or a member of Γ or it follows from predecessors by *modus ponens* or definitional interchange. Consider the sequence $\phi \rightarrow \theta_1, \phi \rightarrow \theta_2, \ldots, \phi \rightarrow \theta_n$. This sequence, though it ends with the sentence $\phi \rightarrow \psi$, may not as it stands be a proof of $\phi \rightarrow \psi$ from the assumptions Γ, but we can convert it into such a proof by adding supplementary steps according to the following directions. If $\theta_i = \phi$, we insert a proof of $\phi \rightarrow \phi$ in place of the step $\phi \rightarrow \theta_i$. If θ_i is an axiom or a member of Γ, we insert the steps

$$\theta_i$$
$$\theta_i \rightarrow (\phi \rightarrow \theta_i)$$

just ahead of the step $\phi \rightarrow \theta_i$. If θ_i follows from predecessors θ_j and θ_k $(= \theta_j \rightarrow \theta_i)$ by *modus ponens,* then just ahead of the step $\phi \rightarrow \theta_i$ we insert the two steps

$$(\phi \rightarrow (\theta_j \rightarrow \theta_i)) \rightarrow ((\phi \rightarrow \theta_j) \rightarrow (\phi \rightarrow \theta_i))$$
$$(\phi \rightarrow \theta_j) \rightarrow (\phi \rightarrow \theta_i).$$

If θ_i follows from a predecessor θ_j by definitional interchange, no additions are necessary, as $\phi \rightarrow \theta_i$ will follow from $\phi \rightarrow \theta_j$ by the same definitional interchange. As may now easily be checked, the sequence resulting from all of these additions is a proof of $\phi \rightarrow \psi$ from the assumptions Γ. Thus $\Gamma \Vdash \phi \rightarrow \psi$.

406. $\Gamma \Vdash \phi$ if and only if $\Gamma \cup \{-\phi\} \Vdash$ 'P & $-P$'.

407. $\Gamma \Vdash \phi$ if and only if, for every sentence ψ, $\Gamma \cup \{-\phi\} \Vdash \psi$.

408. $\Gamma \Vdash \phi$ if and only if ϕ is a consequence of Γ.

The proof is as in Chapter 8, section 3.

EXERCISES

1. Give in full a proof of the sentence

$$(x)(y)(Fxy \to Gxy) \to ((x)(y)Fxy \to (x)(y)Gxy).$$

2. Prove metatheorems 142 through 146.
3. Show that if $\Gamma \Vdash \phi$, then there is a finite set of sentences Δ such that Δ consists exclusively of members of Γ and/or axioms of logic, and ϕ is a tautological consequence of Δ. (Thus, in particular, every valid sentence is a tautological consequence of a finite set of axioms of logic.)

11

FORMALIZED THEORIES

A theory formalized by means of elementary logic is called an *elementary theory*. In the present chapter we consider some important properties of elementary theories, and we illustrate these properties in a series of examples. This is also the place in which to discuss formal rules for introducing definitions, since correctness or incorrectness is a characteristic that a definition possesses only in relation to a given theory. For background and supplementary material in connection with the subject of this chapter the student is advised to read Chapters VI to X, inclusive, of Tarski's *Introduction to Logic*.

1. *Introduction.* Abstractly considered, any group of sentences concerning a given subject matter may be regarded as constituting the assertions or theses of a deductive theory, provided only that one very minimal condition is satisfied: all consequences of theses shall be theses, if they concern the relevant subject matter. If we take the words 'concerning a given subject matter' to amount to 'couched in a given vocabulary', we reach the following more precise definition. *T is a deductive theory formalized in the first order predicate calculus* if and only if T is a pair of sets $\langle \Delta, \Gamma \rangle$, where Δ is a set of non-logical constants of $\mathfrak{L}$ containing at least one predicate of degree ≥ 1 and Γ is a set of sentences of $\mathfrak{L}$ satisfying the conditions:

1) all non-logical constants occurring in members of Γ are members of Δ, and

2) every sentence of T that is a consequence of Γ is a member of Γ

(where a *sentence* (or *formula*) *of* T is a sentence (or formula) all non-logical constants of which belong to Δ), i.e., Γ is *deductively closed*.

When $T = \langle \Delta, \Gamma \rangle$, we call the members of Δ the *non-logical vocabulary* of T and those of Γ the *assertions*, or *theses*, of T. Thus, a deductive theory is uniquely determined by its non-logical vocabulary and its assertions.

Similarly, we shall say that T is a *deductive theory formalized in the first order predicate calculus with identity* (or *with identity and operation symbols*) if and only if the same conditions are satisfied, except that we now refer to $\mathfrak{L}_I$ (or $\mathfrak{L}'$) instead of to $\mathfrak{L}$, so that 'consequence' means 'consequence in $\mathfrak{L}_I$' (or 'consequence in $\mathfrak{L}'$') instead of 'consequence in $\mathfrak{L}$.'* Usually when we use the term 'theory' without qualification we shall be referring to theories formalized in the first order predicate calculus with identity and operation symbols, though most of what we say about such theories can be carried over with little or no change to theories formalized in the predicate calculus with or without identity.

Note that actually a theory is always uniquely determined by the set of its assertions. For suppose that the set of assertions of a theory T is identical with that of a theory T'. Now each non-logical constant in the vocabulary of T occurs in some assertion of T, since all logically valid sentences of T are assertions of T; hence it occurs in the assertions of T' and is part of the vocabulary of T'. Correspondingly, the vocabulary of T' is included in that of T. Thus T and T' have the same vocabulary and assertions, and therefore they are identical.

A theory T is *consistent* if and only if no sentence and its negation are both asserted by T; it is *complete* if and only if, for any sentence of T, either that sentence or its negation is asserted by T. Thus every inconsistent theory is automatically complete.

Another concept useful in the present connection is that of 'independence'; a set of sentences Γ is *independent* if and only if no element ϕ of Γ is a consequence of $\Gamma \sim \{\phi\}$ (for the notation, see Exercise 8, page 42). The question of independence arises most often in relation to sets of axioms (see below); usually we make an effort to select axioms in such a way that no one of them is a consequence of the others. Such an axiom would be theoretically superfluous, though in practice its presence might be very helpful in facilitating deduction of theorems (i.e., assertions). It would perhaps be more natural to define a set of sentences Γ as independent if, for every element ϕ of Γ, neither ϕ nor $-\phi$ is a consequence of $\Gamma \sim \{\phi\}$. But for the cases in which we are most interested, where Γ is consistent, the two definitions come to the same thing.

Let us say further that a set Γ of sentences is *decidable* if there is a step-by-step procedure which, applied to any arbitrary expression ϕ, will decide

* For deductive theories formalized in $\mathfrak{L}_I$ or $\mathfrak{L}'$ it is convenient to drop the condition that Δ shall contain at least one predicate of degree ≥ 1, for trivialization will in these cases be blocked by the presence of the logical predicate 'I_1^2' in theses of every theory.

in a finite number of steps whether or not ϕ belongs to Γ. Note that we do not require the procedure to be known or to be practicable. If Γ contains only finitely many sentences, then Γ is decidable even if the number is very large, for in this case the procedure is simply one of looking through Γ for ϕ. If Γ is infinite, however, it may or may not be decidable. The set of all sentences of $\mathfrak{L}$ is decidable though infinite; the set of all valid sentences of $\mathfrak{L}$, on the other hand, is an infinite set that is not decidable. A theory is said to be decidable if the set of its theses is decidable.

If among the assertions of a given theory T there is a decidable subset of which all the others are consequences, the theory T is *axiomatizable.* In such case the subset in question is called a *set of axioms* for T; and, relative to that set of axioms, T is called an *axiomatized* or *axiomatic theory*.

Axiomatic theories are of particular interest because historically their counterparts in the natural languages have played a very important role in mathematics and even a fairly important role in certain portions of physics. Until recent years every schoolboy was made more or less familiar with some version of Euclid's axiomatic development of geometry. No doubt he soon forgot the proofs, the theorems, and even the axioms, but he seldom lost entirely what may be called 'the general idea': namely, that certain geometrical principles were assumed without proof, and that the problem was to prove other geometrical assertions on the basis of these. The student whose mathematical education goes beyond high school soon comes into contact with numerous other axiom systems—for the integers, for the real numbers, for groups, rings, fields, and a wide variety of further mathematical structures. In physics he finds that various portions of mechanics, too, have been more or less rigorously axiomatized. Indeed, the axiomatic theory is regarded, in some quarters at least, as the ideal form in which one should strive to organize the scientific knowledge of any subject. Whatever scepticism such a view may justifiably call forth, there can be no doubt that the axiomatic method has already been of immense utility in the advancement of mathematical knowledge and in the transmission of that knowledge to students.

Axiomatic theories arise in two different ways, which have been usefully distinguished by Hermes and Markwald.* Sometimes, as was almost certainly the case with Euclidean geometry, the assertions of the theory are characterized more or less adequately before the question of finding axioms comes up; then axiom sets are proposed and are judged on the basis of whether or not the resulting theorems coincide with the assertions (or 'truths') of the theory. Hermes and Markwald call this sort of axiom system a *heteronomous* system. Sometimes, however, the axioms are named first, and the theses of the theory are *defined* as those sentences of the theory that follow from the axioms. In this case the axiom system is

*Hermes, H., and W. Markwald. 'Grundlagen der Mathematik', in *Grundzüge der Mathematik,* ed. H. Behnke, K. Fladt, and W. Süss, Band I, Göttingen, 1958.

called *autonomous.* The division of axiom systems into autonomous and heteronomous is of course a division that is heuristic rather than strictly logical; for, from the standpoint of logic, an axiom system is an axiom system, whether it was discovered after the assertions were already characterized, or whether it is used for such characterization. Also, as the cited authors emphasize, the distinction itself involves a certain amount of idealization, for in practical cases it is frequently difficult to decide whether a given axiom system is to be classified one way or the other. Often autonomous axiom systems arise by omitting or otherwise changing one or more axioms of a heteronomous system; in this way systems of non-Euclidean geometry were obtained from the Euclidean theory.

It is customary to evaluate axiom systems on various counts: consistency, completeness, independence, and categoricity. Consistency is established either semantically (by giving an interpretation under which all the axioms are true) or syntactically (by showing, without reference to interpretations, that there is no sentence ϕ such that both ϕ and $-\phi$ are derivable from the axioms). The importance of consistency is no doubt obvious: from an inconsistent set of axioms every sentence follows, and hence when we go to the trouble of establishing that a given sentence follows from such a set of axioms we do not distinguish it in any interesting way from any other sentence, not even from its own negation.

Consistency is also important in connection with certain quasi-philosophical discussions that arise concerning axiom systems. A common way of replying to such questions as 'what do you mean by 'point'?', 'what do you mean by 'line'?', 'what is a set?', is to say that a point, line, or set is whatever satisfies such and such axioms. Clearly the import of this type of answer depends essentially upon whether or not the axioms are consistent. For most theories having what may loosely be described as 'a relatively large amount of mathematical content', e.g., systems of axiomatic set theory, the problem of establishing consistency has turned out to involve very deep-rooted difficulties.

An axiom system for a given theory is said to be complete just in case the theory itself is complete. Autonomous axiom systems are usually incomplete, as would be expected from the way in which they come into being. In the case of heteronomous systems we are more likely to find completeness, for, with respect to such a system, when someone discovers a sentence that is true (relative to the intended interpretation) but not derivable, an attempt is ordinarily made to remedy this by strengthening the axiom system.

The independence of a set of axioms is most usually demonstrated semantically, by giving for each axiom an interpretation under which it is false but the remaining axioms are true. A set of axioms that is not independent is usually felt to be inelegant, for the derivable axiom is in an obvious sense superfluous even though its presence may greatly facilitate the actual deduction of theorems. Efforts to prove axioms independent have been very fruitful historically. The prime example in this regard is the long struggle to derive Euclid's axiom of parallels, which led eventually to the creation of non-Euclidean geometries. In general, one may say at least that when an axiom system is independent one can much more easily see the contribution of each individual axiom.

An axiom system is *categorical* if all its models are isomorphic, in a sense now to be explained. As mentioned above, when one tries to characterize a notion by means of a set of axioms, it is, of course, essential that the axioms be consistent, i.e., that there be at least one model. Otherwise, we may in effect be saying, e.g., that a point is whatever satisfies such and such axioms, while, unbeknown to us, *nothing* can satisfy those axioms. On the other hand, if the axioms are consistent, we run into another problem: there may be too many models. Our efforts to characterize a notion or notions may fail because we have not been sufficiently restrictive.

Now there is no such thing as an axiom system that has exactly one model, and so the attempt to characterize a model uniquely by giving axioms cannot succeed. Any consistent axiom system will have infinitely many models. The most we can expect, therefore, is that although the models of the axiom system may be numerous they should all be isomorphic, i.e., have the same structure. Given a set of sentences Γ of $\mathfrak{L}'$, we shall say that the interpretations $\mathfrak{I}_1$ and $\mathfrak{I}_2$ are *isomorphic models* of Γ if and only if

(i) $\mathfrak{I}_1$ and $\mathfrak{I}_2$ are models of Γ;

(ii) there is a one-to-one relation between the domains of $\mathfrak{I}_1$ and $\mathfrak{I}_2$, associating with each element $\mathfrak{e}$ of the domain of $\mathfrak{I}_1$ exactly one element $\bar{\mathfrak{e}}$ of the domain of $\mathfrak{I}_2$;

(iii) for each sentential letter ϕ occurring in Γ, $\mathfrak{I}_1(\phi) = \mathfrak{I}_2(\phi)$, i.e., what $\mathfrak{I}_1$ associates with ϕ is identical with what $\mathfrak{I}_2$ associates with ϕ;

(iv) for each individual constant β occurring in Γ, $\mathfrak{I}_2(\beta) = \overline{\mathfrak{I}_1(\beta)}$;

(v) for each n-ary predicate θ occurring in Γ, the n-tuple $\langle \mathfrak{x}_1, \mathfrak{x}_2, \ldots, \mathfrak{x}_n \rangle \in \mathfrak{I}_1(\theta)$ if and only if $\langle \bar{\mathfrak{x}}_1, \bar{\mathfrak{x}}_2, \ldots, \bar{\mathfrak{x}}_n \rangle \in \mathfrak{I}_2(\theta)$, for all elements $\mathfrak{x}_1, \mathfrak{x}_2, \ldots, \mathfrak{x}_n$ of the domain of $\mathfrak{I}_1$;

(vi) for each n-ary operation symbol θ occurring in Γ, $\langle \mathfrak{x}_1, \mathfrak{x}_2, \ldots, \mathfrak{x}_n, \mathfrak{y} \rangle \in \mathfrak{I}_1(\theta)$ if and only if $\langle \bar{\mathfrak{x}}_1, \bar{\mathfrak{x}}_2, \ldots, \bar{\mathfrak{x}}_n, \bar{\mathfrak{y}} \rangle \in \mathfrak{I}_2(\theta)$, for all elements $\mathfrak{x}_1, \mathfrak{x}_2, \ldots, \mathfrak{x}_n, \mathfrak{y}$ of the domain of $\mathfrak{I}_1$.

Categoricity, as thus defined for the first-order language $\mathfrak{L}'$, is a relatively trivial notion. None of the usual axiomatically formulated mathematical theories will be categorical, because any set of sentences of $\mathfrak{L}'$ with an infinite model will have models that are of differing cardinality and hence are not isomorphic. In some of these cases, however, one can establish a so-called representation theorem, to the effect that, although not all the models are isomorphic to one another, each is isomorphic to a model having some special property. For example, while not all models of the theory of groups (see section 4) are isomorphic to one another, every such model is isomorphic to a group of transformations.

2. *Aristotelian syllogistic*. As an example of an axiomatized theory formulated in the first order predicate calculus (without identity) we consider the following theory T_1. The non-logical vocabulary of T_1 consists of four binary predicates 'A^2', 'E^2', 'I^2', 'O^2', and the assertions of T_1 are those of its sentences that are consequences of the following seven axioms:

1. $(x)(y)(z)((Ayz \,\&\, Axy) \rightarrow Axz)$	Barbara
2. $(x)(y)(z)((Eyz \,\&\, Axy) \rightarrow Exz)$	Celarent
3. $(x)(y)(Ixy \rightarrow Iyx)$	Conversion of '*I*'
4. $(x)(y)(Exy \rightarrow Eyx)$	Conversion of '*E*'
5. $(x)(y)(Axy \rightarrow Iyx)$	Conversion of '*A*'
6. $(x)(y)(Exy \leftrightarrow -Ixy)$	Definition of '*E*'
7. $(x)(y)(Oxy \leftrightarrow -Axy)$	Definition of '*O*'

This theory is a formalized version of Aristotle's theory of the syllogism. That it is consistent follows from the fact that axioms 1–7 are all true under any interpretation $\mathfrak{I}$ having as its domain the set of all non-empty subsets of some non-empty set $\mathfrak{m}$, and assigning relations to the four predicates as follows:

$$\begin{aligned} A&: \text{①} \subset \text{②} \\ E&: \text{①} \cap \text{②} = \Lambda \\ I&: \text{①} \cap \text{②} \neq \Lambda \\ O&: \text{①} \not\subset \text{②} \end{aligned}$$

On the other hand, T_1 is not complete, for neither the sentence '$(\exists x)Axx$' nor its negation is an assertion. For, let $\mathfrak{I}'$ be an interpretation having {Aristotle} as domain and assigning relations to the predicates as follows:

$$\begin{aligned} A&: \text{①} \neq \text{②} \\ E&: \text{①} = \text{②} \\ I&: \text{①} \neq \text{②} \\ O&: \text{①} = \text{②} \end{aligned}$$

Then all of the axioms are true under $\mathfrak{I}'$, but '$(\exists x)Axx$' is false. Under the preceding interpretation $\mathfrak{I}$ all of the axioms are true but '$-(\exists x)Axx$' is false. Hence, neither '$(\exists x)Axx$' nor its negation is a consequence of the axioms, and therefore T_1 is incomplete.

The axioms are not independent, either, for obviously axiom 3 can be derived from 4 and 6. Nor are they categorical (thinking here of T_1 as formulated in $\mathfrak{L}'$), since there are models with domains having varying numbers of elements. Even for a given number of elements there are non-isomorphic models. For example, let $\mathfrak{I}_1$ and $\mathfrak{I}_2$ be interpretations having the set {0,1} as domain; let $\mathfrak{I}_1$ make the following assignments:

A: $\{\langle 0,0\rangle, \langle 1,1\rangle, \langle 0,1\rangle\}$
E: Λ
I: $\{\langle 0,0\rangle, \langle 1,1\rangle, \langle 0,1\rangle, \langle 1,0\rangle\}$
O: $\{\langle 1,0\rangle\}$

and $\mathfrak{I}_2$ the following:

A: $\{\langle 0,0\rangle, \langle 1,1\rangle\}$
E: Λ
I: $\{\langle 0,0\rangle, \langle 1,1\rangle, \langle 0,1\rangle, \langle 1,0\rangle\}$
O: $\{\langle 0,1\rangle, \langle 1,0\rangle\}$

If $\mathfrak{I}_1$ and $\mathfrak{I}_2$ were isomorphic, we would have $\langle 0,1\rangle \in \mathfrak{I}_1$('$A$') if and only if $\langle \bar{0},\bar{1}\rangle \in \mathfrak{I}_2$('$A$'); but 0 and $\bar{1}$ must be distinct, since for isomorphism of two interpretations the correspondence between the domains is one-to-one, and yet no couple with distinct terms belongs to $\mathfrak{I}_2$('A').

The intended interpretations of T_1 are those of the sort we described in showing the axioms consistent. Suppose, for instance, that $\mathfrak{m}$ is the set of all living things. Let 'α', 'β', 'γ' be (metalinguistic) variables ranging over the non-empty subsets of $\mathfrak{m}$. Then we find that axioms 1–7 represent, respectively, the following:

1′. For all α,β,γ: if all β are γ, and all α are β, then all α are γ.
2′. For all α,β,γ: if no β are γ, and all α are β, then no α are γ.
3′. For all α,β: if some α are β, then some β are α.
4′. For all α,β: if no α are β, then no β are α.
5′. For all α,β: if all α are β, then some β are α.
6′. For all α,β: no α are β if and only if it is not the case that some α are β.
7′. For all α,β: some α are not β if and only if not all α are β.

Each of the foregoing is a logical principle that may, with certain qualifications, be ascribed to Aristotle (see Chapter 12, section 1). Statements 1′ and 2′ are the syllogistic moods called 'Barbara' and 'Celarent'; 3′, 4′ and 5′ are the so-called laws of conversion; 6′ and 7′ are part of the information constituting the Aristotelian Square of Opposition. All are used by Aristotle in his derivation of the remaining valid syllogistic moods from Barbara and Celarent. It is possible to reproduce in our natural deduction system Aristotle's own derivations.

For the sake of brevity, however, let us first introduce a few conventions for telescoping proofs. The citation 'US′' will mean 'by repeated application of US', and similarly for other rules. Also, the following rule will be of considerable help.

TH Any instance* of an axiom or previously proved theorem may be entered on a line, with the empty set of premise-numbers; more

*Let us say, in this connection, that ϕ is an *instance* of ψ if and only if ϕ can be obtained from ψ by repeated application of US.

generally, ψ may be entered on a line if $\phi_1, \ldots, \phi_n$ appear on earlier lines and ψ is a tautological consequence of $\{\phi_1, \ldots, \phi_n\}$ and an instance of an axiom or previously proved theorem; as premise-numbers of the new line take all premise-numbers of those earlier lines.

Note that in connection with premise-numbers this rule allows us to dispense with reference to instances of axioms or previously proved theorems. Thus, in the following derivations lines may depend upon axioms 1–7 as well as upon the premises explicitly indicated. Since all axioms of T_1 are also theorems of T_1, we begin our numbering with '8'.

8. $(x)(y)(z)((Ezy \,\&\, Axy) \rightarrow Exz)$ Cesare

{1}	(1)	$Ecb \,\&\, Aab$	P
{1}	(2)	$Ebc \,\&\, Aab$	(1),4,TH
{1}	(3)	Eac	(2),2,TH
Λ	(4)	$(Ecb \,\&\, Aab) \rightarrow Eac$	(1),(3),C
Λ	(5)	$(x)(y)(z)((Ezy \,\&\, Axy) \rightarrow Exz)$	(4),UG′

9. $(x)(y)(z)((Azy \,\&\, Exy) \rightarrow Exz)$ Camestres

{1}	(1)	$Acb \,\&\, Eab$	P
{1}	(2)	$Eba \,\&\, Acb$	(1),4,TH
{1}	(3)	Eca	(2),2,TH
{1}	(4)	Eac	(3),4,TH
Λ	(5)	$(Acb \,\&\, Eab) \rightarrow Eac$	(1),(4),C
Λ	(6)	$(x)(y)(z)((Azy \,\&\, Exy) \rightarrow Exz)$	(5),UG′

10. $(x)(y)(z)((Ayz \,\&\, Ixy) \rightarrow Ixz)$ Darii

{1}	(1)	$Abc \,\&\, Iab$	P
{2}	(2)	$-Iac$	P
{1,2}	(3)	$Abc \,\&\, Eac$	(1),(2),6,TH
{1,2}	(4)	Eab	(3),9,TH
{1}	(5)	$-Iac \rightarrow Eab$	(2),(4),C
{1}	(6)	Iac	(1),(5),6,TH
Λ	(7)	$(Abc \,\&\, Iab) \rightarrow Iac$	(1),(6),C
Λ	(8)	$(x)(y)(z)((Ayz \,\&\, Ixy) \rightarrow Ixz)$	(7),UG′

11. $(x)(y)(z)((Eyz \,\&\, Ixy) \rightarrow Oxz)$ Ferio
(Proof analogous to that of theorem 10, except that 7, 8, and 6 are used.)
12. $(x)(y)(z)((Ezy \,\&\, Ixy) \rightarrow Oxz)$ Festino
(Proof analogous to that of theorem 8, except that 4 and 11 are used).
13. $(x)(y)(z)((Azy \,\&\, Oxy) \rightarrow Oxz)$ Baroco
(Proof analogous to that of theorem 10, except that 7 and 1 are used).
14. $(x)(y)(z)((Ayz \,\&\, Ayx) \rightarrow Ixz)$ Darapti 5,10
15. $(x)(y)(z)((Eyz \,\&\, Ayx) \rightarrow Oxz)$ Felapton 5,11
16. $(x)(y)(z)((Iyz \,\&\, Ayx) \rightarrow Ixz)$ Disamis 3,10

17. $(x)(y)(z)((Ayz \,\&\, Iyx) \rightarrow Ixz)$ Datisi 3,10
18. $(x)(y)(z)((Oyz \,\&\, Ayx) \rightarrow Oxz)$ Bocardo
(Proof analogous to that of theorem 10, except that 7 and 1 are used)
19. $(x)(y)(z)((Eyz \,\&\, Iyx) \rightarrow Oxz)$ Ferison 3,11

The student may be interested in comparing the foregoing proofs with the Aristotelian statements quoted in Chapter 12, section 1.

3. *The theory of betweenness*. Let us next consider an example of an axiomatized theory formulated in the first order predicate calculus with identity (but without operation symbols). This theory will be called 'T_2'. Its non-logical vocabulary consists of the single predicate 'B^3', and thus its sentences are all those sentences of $\mathfrak{L}_I$ in which no non-logical constant other than 'B^3' occurs. (Remember that in the language $\mathfrak{L}_I$ the predicate 'I_1^2', for identity, is classified as a logical constant.) The assertions of T_2 may be defined as those of its sentences that are true under every interpretation $\mathfrak{I}$ satisfying the two conditions:

(i) The domain of $\mathfrak{I}$ is the set of all points on some (Euclidean) straight line, and

(ii) $\mathfrak{I}$ assigns to 'B^3' the relation of betweenness (as given by the predicate '② is between ① and ③') among points on this straight line. 'Between' has its usual meaning, except that we regard each point as being between itself and any other point, and between itself and itself.

It will be noticed that the assertions of T_2, as thus specified, satisfy the condition that all consequences of assertions are again assertions, for the consequences of a set are true under all interpretations under which all members of the set are true. Further, since no sentence and its negation are both true under any interpretation, the theory T_2 is consistent. Indeed, if a sentence of T_2 is true under one interpretation of the sort described above, it is true under all such interpretations; therefore, for every sentence of T_2, either that sentence or its negation is asserted by T_2. Thus T_2 is also complete.

Now it turns out that T_2 is axiomatizable by means of the following seven axioms:

1. $(x)(y)(Bxyx \rightarrow x = y)$
2. $(x)(y)(z)(u)((Bxyu \,\&\, Byzu) \rightarrow Bxyz)$
3. $(x)(y)(z)(u)(((Bxyz \,\&\, Byzu) \,\&\, y \neq z) \rightarrow Bxyu)$
4. $(x)(y)(z)((Bxyz \vee Bxzy) \vee Bzxy)$
5. $(\exists x)(\exists y)x \neq y$
6. $(x)(y)(x \neq y \rightarrow (\exists z)(Bxyz \,\&\, y \neq z))$
7. $(x)(y)(x \neq y \rightarrow (\exists z)(Bxzy \,\&\, (z \neq x \,\&\, z \neq y)))$

In writing out derivations we make use of the same abbreviating conventions employed in connection with T_1; for the language $\mathfrak{L}_I$, however, it will be convenient to strengthen rule TH (see pages 189–90) by inserting the phrase 'or derivable by an application of rule I from' immediately after the phrase 'a tautological consequence of'.

8. $(x)(y)(Bxxy \,\&\, Bxyy)$

Λ	(1) $(Baba \vee Baab) \vee Baab$	TH,4
$\{2\}$	(2) $Baba$	P
$\{2\}$	(3) $a = b$	(2),TH,1
$\{2\}$	(4) $Baab$	(2),(3),I
Λ	(5) $Baba \rightarrow Baab$	
Λ	(6) $Baab$	
Λ	(7) $(Babb \vee Babb) \vee Bbab$	TH,4
$\{8\}$	(8) $Bbab$	P
$\{8\}$	(9) $b = a$	(8),TH,1
$\{8\}$	(10) $Babb$	(8),(9),I
Λ	(11) $Bbab \rightarrow Babb$	
Λ	(12) $Babb$	
Λ	(13) $Baab \,\&\, Babb$	
Λ	(14) $(x)(y)(Bxxy \,\&\, Bxyy)$	(13),UG′

It will be observed that many sentences of the foregoing derivation are not sentences of the theory T_2. This may at first seem to be an unwelcome consequence of the way in which we have defined 'theory' or of the way in which we have set up inference rules for the predicate calculus. But a little reflection will show that in constructing proofs in ordinary theories—e.g., Euclidean geometry—we assert many things that are not strictly geometrical assertions, or on which we would expect a geometrical theory to 'take sides'. For instance, one may say 'let *a* and *b* be two intersecting lines'. One would not feel that a geometry was incomplete merely because it contained neither this sentence nor its negation.

9. $(x)(y)(z)(Bxyz \rightarrow Bzyx)$

$\{1\}$	(1) $a = b$	P
$\{2\}$	(2) $b = c$	P
Λ	(3) $Bcbb$	TH,8
$\{1\}$	(4) $Bcba$	TH,(1),(3)
Λ	(5) $Bcca$	TH,8
$\{2\}$	(6) $Bcba$	TH,(2),(5)
Λ	(7) $(a = b \vee b = c) \rightarrow Bcba$	C′
Λ	(8) $((Babc \,\&\, Bbca) \,\&\, b \neq c) \rightarrow Baba$	3,US′
Λ	(9) $(Babc \,\&\, Bbac) \rightarrow Baba$	2,US′
Λ	(10) $(Bbac \vee Bbca) \vee Bcba$	4,US′
Λ	(11) $Babc \rightarrow Bcba$	1,(7)–(10),TH
Λ	(12) $(x)(y)(z)(Bxyz \rightarrow Bzyx)$	UG′,(11)

10. $(x)(y)(z)((Bxyz \,\&\, Byzx) \to y = z)$

Λ	(1) $Babc \to Bcba$	9,TH
Λ	(2) $(Bbca \,\&\, Bcba) \to Bbcb$	2,US′
Λ	(3) $(Babc \,\&\, Bbca) \to b = c$	1,(1),(2),TH
Λ	(4) $(x)(y)(z)((Bxyz \,\&\, Byzx) \to y = z)$	(3),UG′

11. $(x)(y)(z)(u)(((Bxyz \,\&\, Byzu) \,\&\, y \neq z) \to Bxzu)$ 3,9
12. $(x)(y)(z)(u)((Bxyu \,\&\, Byzu) \to Bxzu)$ 2,11
13. $(x)(y)(z)(u)((Bxyu \,\&\, Byzu) \to (Bxyz \,\&\, Bxzu))$ 2,9,12
14. $(x)(y)(z)(u)((Bxyu \,\&\, Bxzu) \to ((Bxyz \,\&\, Byzu) \vee (Bxzy \,\&\, Bzyu)))$ 3,4,8,9,10,13
15. $(x)(y)(z)(u)(((Bxyz \,\&\, Bxyu) \,\&\, x \neq y) \to ((Bxzu \,\&\, Byzu) \vee (Bxuz \,\&\, Byuz)))$ 9,10,13
16. $(\exists x)(\exists y)(\exists z)((x \neq y \,\&\, x \neq z) \,\&\, y \neq z)$
17. $(\exists x)(\exists y)(\exists z)(\exists u)(((x \neq y \,\&\, x \neq z) \,\&\, (x \neq u \,\&\, y \neq z)) \,\&\, (y \neq u \,\&\, z \neq u))$
18. $(x)(y)(z)(u)(Bxyz \to (Bxyu \vee Buyz))$ 2,3,4,9
19. $(x)(y)(z)(u)(v)(((Bxzy \,\&\, Bxuy) \,\&\, Bzvu) \to Bxvy)$
20. $(x)(y)(z)(u)(((Bxyz \,\&\, Byzu) \,\&\, Bzux) \to (y = z \vee z = u))$

We have seen earlier that the theory T_2 is consistent. The question now arises whether our axiom system is independent. To show this we must find for each axiom an interpretation under which that axiom is false and the other six are true. Thus, for example, we can establish that axiom 1 is not derivable from axioms 2–7 by giving an interpretation $\mathfrak{I}$ as follows:

$\mathfrak{D}$: the set of all points on a straight line $\mathfrak{l}$
B^3: ① is a point on $\mathfrak{l}$ and ② is a point on $\mathfrak{l}$ and ③ is a point on $\mathfrak{l}$.

Under this sort of interpretation axiom 1 is false but axioms 2–7 are true; therefore, axiom 1 is not a consequence of axioms 2–7. For another example, consider an interpretation $\mathfrak{I}'$ described as follows:

$\mathfrak{D}$: the set of all points in a plane $\mathfrak{p}$
B^3: ①, ②, ③ lie on a straight line in $\mathfrak{p}$, and ② is between ① and ③.

Under this sort of interpretation axiom 4 is false but the remaining axioms are true; therefore, axiom 4 is not a consequence of the others. We leave to the reader the task of finding the five other interpretations that are needed in order to demonstrate that the axiom system is independent.

4. *Groups; Boolean algebras*. When we have not only identity but also operation symbols at our disposal, it is possible to formalize in a perspic-

uous way a very wide variety of theories. For example, consider the theory T_3, the *theory of groups,* formalized in the language $\mathfrak{L}'$. The non-logical vocabulary of T_3 will consist of the binary operation symbol 'f^2' (for so-called group addition), the singulary operation symbol 'f^1' (for group complementation), and the individual constant 'e' (for the identity element). Thus, the sentences of T_3 are all sentences of $\mathfrak{L}'$ containing no non-logical constants other than the three just mentioned. The assertions are specified as all consequences of certain axioms. These are more advantageously written with the help of two conventions: in sentences of T^3 we may write

$$(\tau + \tau') \quad \text{for} \quad f^2\tau\tau'$$

and

$$\bar{\tau} \quad \text{for} \quad f^1\tau,$$

where τ,τ' are any terms or the results of previous application of these conventions. Outermost parentheses will be dropped from a term when there is no danger of ambiguity. The axioms of T_3 are then three in number:

1. $(x)(y)(z)\, x + (y + z) = (x + y) + z$
2. $(x)\, x + e = x$
3. $(x)\, x + \bar{x} = e$

Obviously the theory is consistent, since the three axioms are true under an interpretation $\mathfrak{I}$ having the integers (positive, negative, and zero) as its domain and making the following assignments to the non-logical vocabulary of the theory:

f^2: ① + ② (in the usual sense of '+')
f^1: 0 − ① (in the usual sense of '−')
e: 0

The theory is not complete, for neither the sentence '$(x)x = e$' nor its negation is a consequence of the axioms, as may easily be checked by constructing appropriate interpretations. In the same way it is easy to show that the axioms are independent. They are of course not categorical, since they possess models with domains having different numbers of elements.

In deriving theorems we must be careful not to generalize upon the individual constant 'e', which appears in the second and third axioms and hence may occur in a premise of a given line even though, through application of rule TH, that premise is not explicitly indicated. Among the theorems are the following:

4. $(x)(y)(z)(x + z = y + z \rightarrow x = y)$

{1}	(1) $a + c = b + c$	P
{1}	(2) $(a + c) + \bar{c} = (b + c) + \bar{c}$	(1),TH
{1}	(3) $a + (c + \bar{c}) = b + (c + \bar{c})$	(2),TH,1

$\{1\}$ (4) $a + e = b + e$ (3),TH,3
$\{1\}$ (5) $a = b$ (4),TH,2
Λ (6) $a + c = b + c \rightarrow a = b$ (1),(5),C
Λ (7) $(x)(y)(z)(x + z = y + z \rightarrow x = y)$ (6),UG′

5. $(x)\, x + e = e + x$

Λ (1) $e + (a + \bar{a}) = (e + a) + \bar{a}$ TH,1
Λ (2) $e + e = (e + a) + \bar{a}$ (1),TH,3
Λ (3) $e = (e + a) + \bar{a}$ (2),TH,2
Λ (4) $a + \bar{a} = (e + a) + \bar{a}$ (3),TH,3
Λ (5) $a = e + a$ (4),TH,4
Λ (6) $a + e = a$ TH,2
Λ (7) $a + e = e + a$ (5),(6),I
Λ (8) $(x)\, x + e = e + x$ (7),UG

6. $(y)((x)x + y = x \rightarrow y = e)$

$\{1\}$ (1) $(x)\, x + a = x$ P
$\{1\}$ (2) $e + a = e$ (1),US
$\{1\}$ (3) $a + e = e$ (2),TH,5
$\{1\}$ (4) $a = e$ (3),TH,2
Λ (5) $(x)x + a = x \rightarrow a = e$ (1),(4),C
Λ (6) $(y)((x)x + y = x \rightarrow y = e)$ (5),UG′

7. $(x)\, x + \bar{x} = \bar{x} + x$

Λ (1) $\bar{a} + (a + \bar{a}) = (\bar{a} + a) + \bar{a}$ TH,1
Λ (2) $\bar{a} + e = (\bar{a} + a) + \bar{a}$ (1),TH,3
Λ (3) $e + \bar{a} = (\bar{a} + a) + \bar{a}$ (2),TH,5
Λ (4) $e = \bar{a} + a$ (3),TH,4
Λ (5) $a + \bar{a} = \bar{a} + a$ (4),TH,3
Λ (6) $(x)\, x + \bar{x} = \bar{x} + x$ (5),UG

8. $(x)(y)(z)(z + x = z + y \rightarrow x = y)$

$\{1\}$ (1) $c + a = c + b$ P
$\{1\}$ (2) $\bar{c} + (c + a) = \bar{c} + (c + b)$ (1),TH
$\{1\}$ (3) $(\bar{c} + c) + a = (\bar{c} + c) + b$ (2),TH,1
$\{1\}$ (4) $(c + \bar{c}) + a = (c + \bar{c}) + b$ (3),TH,7
$\{1\}$ (5) $e + a = e + b$ (4),TH,3
$\{1\}$ (6) $a + e = b + e$ (5),TH,5
$\{1\}$ (7) $a = b$ (6),TH,2
Λ (8) $c + a = c + b \rightarrow a = b$ (1),(7),C
Λ (9) $(x)(y)(z)(z + x = z + y \rightarrow x = y)$ 8,UG′

9. $(x)(y)(x + y = e \rightarrow y = \bar{x})$
10. $(x)\, \bar{\bar{x}} = x$
11. $(x)(y)(\exists z)(x = y + z \,\&\, (w)(x = y + w \rightarrow w = z))$
12. $(x)(z)(\exists y)(x = y + z \,\&\, (w)(x = w + z \rightarrow w = y))$

As a final example of a formalized theory we consider T_4, the theory of *Boolean algebras*. This theory may be formulated in the first order predicate calculus with identity and operation symbols, using as non-logical vocabulary the five signs 'f^1', 'f^2', 'g^2', 'n', and 'e'. If we write

$$
\begin{array}{lll}
(\tau \cup \tau') & \text{for} & f^2\tau\tau' \\
(\tau \cap \tau') & \text{for} & g^2\tau\tau' \\
\bar{\tau} & \text{for} & f^1\tau \\
\text{`0'} & \text{for} & \text{`}n\text{'} \\
\text{`1'} & \text{for} & \text{`}e\text{'},
\end{array}
$$

where τ and τ' are either terms or result from previous applications of these conventions, and again drop outside parentheses when no ambiguity threatens, our axioms look as follows:

1. $(x)(y)\, x \cap y = y \cap x$
2. $(x)(y)\, x \cup y = y \cup x$ } (Commutative laws)
3. $(x)(y)(z)\, x \cap (y \cap z) = (x \cap y) \cap z$
4. $(x)(y)(z)\, x \cup (y \cup z) = (x \cup y) \cup z$ } (Associative laws)
5. $(x)(y)\, x \cap (x \cup y) = x$
6. $(x)(y)\, x \cup (x \cap y) = x$ } (Absorption laws)
7. $(x)(y)(z)\, x \cap (y \cup z) = (x \cap y) \cup (x \cap z)$
8. $(x)(y)(z)\, x \cup (y \cap z) = (x \cup y) \cap (x \cup z)$ } (Distributive laws)
9. $(x)\, x \cap \bar{x} = 0$
10. $(x)\, x \cup \bar{x} = 1$ } (Laws of the complement)
11. $0 \neq 1$

An interpretation $\mathfrak{I}$ qualifies as an intended interpretation of T_4 if its domain is a non-empty set $\mathfrak{F}$ of subsets of some non-empty set $\mathfrak{m}$, and it assigns union, intersection, complementation (relative to $\mathfrak{m}$), the empty set, and $\mathfrak{m}$, respectively, to 'f^2', 'g^2', 'f^1', 'n', and 'e'; $\mathfrak{F}$ must satisfy the added condition, of course, that the union and intersection of every pair of sets in $\mathfrak{F}$ shall be in $\mathfrak{F}$, and similarly for complementation relative to $\mathfrak{m}$. The theses of T_4 are those sentences of T_4 that are true under all the intended interpretations. T_4 is consistent but not complete. It is axiomatizable in many ways. The foregoing set of eleven axioms is not the simplest set available, but it makes the derivations relatively easy. The theorems are of course already familiar from Chapter 2.

12. $(x)\, x \cap x = x$
13. $(x)\, x \cup x = x$
14. $(x)\, x \cap 0 = 0$
15. $(x)\, x \cup 1 = 1$
16. $(x)\, x \cup 0 = x$
17. $(x)\, x \cap 1 = x$
18. $(x)(y)(x \cup y = y \rightarrow x \cap y = x)$
19. $(x)(y)(x \cap y = y \rightarrow x \cup y = x)$
20. $0 = 1$

21. $1 = 0$
22. $(x)(y)\,\overline{x \cup y} = \bar{x} \cap \bar{y}$
23. $(x)(y)\,\overline{x \cap y} = \bar{x} \cup \bar{y}$
24. $(x)(y)(x \cap y) \cup (\bar{x} \cap y) = (x \cup \bar{y}) \cap (\bar{x} \cup y)$
25. $(x)(y)(x \cup y) \cap (\bar{x} \cup \bar{y}) = (x \cap \bar{y}) \cup (\bar{x} \cap y)$
26. $(x)\,\bar{\bar{x}} = x$
27. $(x)\,\bar{x} \neq x$
28. $(x)(y)\,x = \bar{y} \leftrightarrow y = \bar{x}$

One obtains the *dual* of a sentence of T_4 by interchange of '$\cap$' and '$\cup$', and of '0' and '1'. Thus axioms 2, 4, 6, 8, and 10 are, respectively, the duals of axioms 1, 3, 5, 7, and 9; axiom 11 is equivalent to its own dual, '$1 \neq 0$'. If throughout the derivation of a theorem we make the interchange indicated, the result will be a derivation of the dual of that theorem. Hence, the dual of a theorem is again a theorem.

5. *Definitions.* In the presentation of a formalized theory one normally employs definitions that serve to introduce notation that does not belong to the vocabulary of the theory but may improve readability of the formulas and render their content more clear. Such a definition may occur in either of two forms: (*a*) as a metalinguistic sentence asserting that a given symbol of the object language means the same as (or abbreviates, or may always be interchanged with) some other object language expression, or (*b*) as an object language sentence of a certain form (usually an identity or biconditional, or a generalized identity or biconditional, with the new symbol appearing on the left side alone or in a simple context). Thus, either of the following expressions could be regarded as a definition of the mathematical symbol '2':

'2' is an abbreviation of '1 + 1'

or

$$2 = 1 + 1.$$

When the defined symbol is of the nature of a predicate or an operation symbol, the definition is ordinarily formulated with the help of variables. For example: for any terms α, β

$$\alpha = \beta \text{ stands for } I^2_1\alpha\beta$$

or

$$(x)(y)(x = y \leftrightarrow I^2_1xy).$$

In general, a definition of type (*a*) will contain a name of the defined symbol, and the variables, if any, that occur in it will be variables of the metalanguage, while a definition of type (*b*) will contain the defined symbol itself together with other expressions of the object language.

We have used both kinds of definition in the present book. In an explicit consideration of the topic, however, it is wise to focus attention upon definitions of type (*b*), since this approach makes the various difficulties and dangers more obvious and permits us to formulate relatively simple rules for avoiding them.

That definitions can be dangerous may come as a surprise to the beginner, especially if he is accustomed to think of them as stated in form (*a*). He may share the attitude of Humpty Dumpty ('when *I* use a word it means what I wish it to mean, neither more nor less'), and from this point of view he may find it hard to see how any difficulty could arise merely from introducing a convention to the effect that a certain new symbol shall stand for a certain string of old ones. But it is not hard to show that such confidence is misplaced. Suppose, to use a standard example, that for the integers it is proposed to define division in terms of multiplication as follows:

$$(x)(y)(z)(\frac{x}{y} = z \leftrightarrow x = yz).$$

From this we can obtain at once the consequences '$\frac{0}{0} = 1$' and '$\frac{0}{0} = 2$', whence also '$1 = 2$', which flatly contradicts a received thesis of the arithmetic of integers. Thus the definition leads to a contradiction. Note that the similarly constructed definition of subtraction in terms of addition would cause no such trouble:

$$(x)(y)(z)(x - y = z \leftrightarrow x = y + z).$$

Therefore the difficulty stems not only from the form of these definitions but depends as well upon the content of the theory to which they are added.

When a definition leads to contradiction, its addition to a previously consistent theory results, of course, in a very large increase in the number of theses. This runs counter to our intuitive demands on a definition, which include that it should lead to *no* essential increase in the set of theses, i.e., that the only new theses resulting from a definition should be such as contain the newly introduced symbol. Sometimes a definition will fail to satisfy this demand even though its addition to a theory does not take us all the way to a contradiction. For example, suppose that to the theory of groups we add as a definition of the binary operation symbol 'g' the sentence:

$$(x)(y)(z)(gxy = z \leftrightarrow x = y).$$

From this we get various new theorems in which 'g' does not occur, including, e.g., the sentence '$(x)(y)\, x = y$'. Thus, although the theory is still consistent, a radical alteration in its content has been made, since now its only models are such as contain just one element in their domains. Again, note that the similar definition

$$(x)(y)(z)(hx = y \leftrightarrow x + y = e)$$

leads to no such difficulties in the theory of groups. A definition that generates new theorems in which the defined symbol does not occur is called *creative;* it seems natural to require of a satisfactory definition that it not be creative.

Another intuitive demand on definitions is that they should make the defined symbols *eliminable.* Whatever can be stated with the help of the defined symbol should be stateable without it, simply by replacing it at every occurrence by the expression it abbreviates. This demand is behind the resistance to 'circular' definitions; if the defined symbol occurs not only on the left of the definition but also on the right, eliminability (at least in the usual way) is thwarted. Other structural defects in a definition can lead to the same result. Consider, for example, the following definition of a quaternary predicate '*R*':

$$(x)(y)(z)(Rxyzx \leftrightarrow x + yz > z).$$

Such a definition only allows us to eliminate an occurrence of '*R*' if its first and fourth arguments are identical; to a formula like '*Rabcd*' it cannot be applied at all.

In order to draw up exact rules for the construction of formally correct definitions, we need to state our two criteria somewhat more sharply. Preparatory to doing this we observe that there are two ways in which, most characteristically, a definition comes under consideration in connection with a formalized theory. In presenting the theory one may wish at a certain point to introduce a new symbol. The motivation may be merely to save space or to increase readability in some other way, or it may also involve introducing notation that is more familiar or has some other heuristic advantage, as for example if one added

$$(x)(y)(x > y \leftrightarrow -(x = y \vee x < y))$$

as a definition to a theory appropriately formulated in terms of '$<$'. Or, within a given theory one may be interested in the question whether a certain constant is definable by means of the remaining constants, i.e., whether a formally correct definition of that constant is included among the theses of the theory. It is easily seen that in either case two theories are involved, a larger and a smaller. The vocabulary of the larger is that of the smaller plus the defined symbol, and the theses of the larger are those of the smaller plus the definition and all consequences of these that can be formulated in the larger vocabulary.

Now let T be a theory, with non-logical vocabulary Δ and theses Γ. Further, let θ be a non-logical constant that does not belong to Δ, and let ϕ be a sentence of $\mathfrak{L}'$ containing θ and otherwise formulated solely in terms of the vocabulary Δ. If we add ϕ to the theses of T we get a new theory T', the non-logical vocabulary of which is $\Delta \cup \{\theta\}$, and the theses of which are all sentences of T' that are consequences of $\Gamma \cup \{\phi\}$. We shall say that ϕ, considered as a definition of θ relative to T, satisfies the criterion of eliminability if and only if, for every formula ψ of T', there is a formula χ of T' such that all closures of $\psi \leftrightarrow \chi$ are theses of T' and χ does not contain θ. Similarly, we say that ϕ, considered as a definition of θ relative to T, satisfies the criterion of non-creativity if and only if every thesis of T' that does not contain θ is a thesis of T.

It is possible to specify certain types of formula that have the desired properties. Let us first do this with reference to the language $\mathfrak{L}_{\mathrm{I}}$. If θ is an n-ary predicate, then any sentence ϕ of the form

$$(\alpha_1) \ldots (\alpha_n)(\theta\alpha_1 \ldots \alpha_n \leftrightarrow \omega),$$

where $\alpha_1, \ldots, \alpha_n$ are distinct variables, and where ω is a formula of T having free occurrences of the variables $\alpha_1, \ldots, \alpha_n$ and of no others, will satisfy the two criteria mentioned. Such a sentence ϕ will satisfy the eliminability criterion, for if ψ is a formula of T', then, by using ϕ to make replacements in the right side of closures of $\psi \leftrightarrow \psi$ (which of course are theses of T'), we obtain as theses all closures of a formula $\psi \leftrightarrow \chi$, where χ contains no occurrence of θ. Also, ϕ will satisfy the criterion of non-creativity. For suppose that the sentence μ is a thesis of T' not containing θ. Then μ is a consequence of $\Gamma \cup \{\phi\}$, and by the completeness theorem it is a consequence of $\Gamma' \cup \{\phi\}$, for some finite subset Γ' of Γ. Let ν be a conjunction of the elements of Γ'. By the deduction theorem,

$$\phi \rightarrow (\nu \rightarrow \mu)$$

is a theorem of logic. The only occurrence of θ in this theorem is the occurrence in ϕ. By making an appropriate substitution for θ we obtain another theorem of logic

$$\phi' \rightarrow (\nu \rightarrow \mu)$$

where ϕ' is a closure of $\omega \leftrightarrow \omega$ and hence is itself a theorem of logic. Therefore

$$\nu \rightarrow \mu$$

is a theorem of logic, and since μ is a sentence of T and follows from the thesis ν of T, μ is a thesis of T.

If θ is a sentential letter, then any sentence of $\mathfrak{L}_{\mathrm{I}}$

$$\theta \leftrightarrow \omega,$$

where ω is a sentence of T, will satisfy the two criteria. The argument is the same as for the case in which θ is an n-ary predicate.

If θ is an individual constant, then any sentence of $\mathfrak{L}_{\mathrm{I}}$ of the form

$$(\alpha)(\alpha = \theta \leftrightarrow \omega),$$

where ω is a formula of T having α as its only free variable, will satisfy our two criteria provided that a sentence

$$(\exists\beta)(\alpha)(\alpha = \beta \leftrightarrow \omega),$$

where β is a variable distinct from α, is a thesis of T. (This sentence says that exactly one object in the domain satisfies ω). To see that such a definition will satisfy the criterion of eliminability, suppose that θ occurs in some formula ψ of T', let α be a variable not occurring in ψ, and let ψ' be like ψ except for having occurrences of α wherever ψ has occurrences of θ. Then every closure of

$$\psi \leftrightarrow (\exists\alpha)(\alpha = \theta \,\&\, \psi')$$

is a theorem of logic and thus a thesis of T'. The definition may now be used to replace the component $\alpha = \theta$ of the right side of these closures, thus giving rise to a formula χ of T such that χ does not contain θ and all closures of $\psi \leftrightarrow \chi$ are theses of T'. Next, to show non-creativity suppose that μ is a thesis of T' not containing θ. As in the case of predicates, we again have the result that

$$(\alpha)(\alpha = \theta \leftrightarrow \omega) \rightarrow (\nu \rightarrow \mu)$$

is a theorem of logic, where ν is a conjunction of theses of T (and hence itself a thesis of T). Thus, if β is a variable distinct from α,

$$(\beta)((\alpha)(\alpha = \beta \leftrightarrow \omega) \rightarrow (\nu \rightarrow \mu))$$

is a theorem of logic, whence

$$(\exists\beta)(\alpha)(\alpha = \beta \leftrightarrow \omega) \rightarrow (\nu \rightarrow \mu)$$

is also a theorem. But $(\exists\beta)(\alpha)(\alpha = \beta \leftrightarrow \omega)$ and ν are theses of T; therefore, μ is also a thesis of T.

We have thus shown how to give satisfactory definitions of the various kinds of non-logical constants that appear in the language $\mathfrak{L}_{\mathrm{I}}$. In the language $\mathfrak{L}'$ the same kinds of sentences, under the same conditions, will serve as definitions of n-ary predicates, sentential letters, and individual constants. There are in addition operation symbols of degree greater than 0. Using the same terminology as before, except that we now regard our theories as formulated in the language $\mathfrak{L}'$ instead of in $\mathfrak{L}_{\mathrm{I}}$, we may say that any closure ϕ of

$$(\alpha)(\alpha = \theta\gamma_1 \ldots \gamma_n \leftrightarrow \omega)$$

where θ is an n-ary operation symbol not occurring in ω, and where $\alpha, \gamma_1, \ldots \gamma_n$ are distinct variables and constitute all variables occurring free in ω, will satisfy the two criteria, provided again that a sentence

$$(\gamma_1) \ldots (\gamma_n)(\exists\beta)(\alpha)(\alpha = \beta \leftrightarrow \omega),$$

where β is a variable distinct from $\alpha, \gamma_1, \ldots, \gamma_n$, is a thesis of T.

All of this may now be summarized in the following rules for the construction of *formally correct definitions*. Let T be a theory formulated in $\mathfrak{L}'$, and suppose that θ is a non-logical constant which does not occur in T but for which it is desired to give a formally correct definition relative to T.

1. If the constant θ is an n-ary predicate, let the definition be a sentence of the form

$$(\alpha_1) \ldots (\alpha_n)(\theta\alpha_1 \ldots \alpha_n \leftrightarrow \omega),$$

where ω is a formula of T and $\alpha_1, \ldots, \alpha_n$ are distinct and are all the variables occurring free in ω.

2. If the constant θ is a sentential letter, let the definition be a sentence

$$\theta \leftrightarrow \omega,$$

where ω is a sentence of T.

3. If the constant θ is an individual constant, let the definition be a sentence

$$(\alpha)(\alpha = \theta \leftrightarrow \omega),$$

where ω is a formula of T and having α as its only free variable, and the corresponding sentence

$$(\exists\beta)(\alpha)(\alpha = \beta \leftrightarrow \omega),$$

with β a variable distinct from α, is a thesis of T.

4. Finally, if the constant θ is an n-ary operation symbol ($n > 0$), let the definition be a sentence of the form

$$(\gamma_1) \ldots (\gamma_n)(\alpha)(\alpha = \theta\gamma_1 \ldots \gamma_n \leftrightarrow \omega),$$

where $\alpha, \gamma_1, \ldots, \gamma_n$ are distinct and are all the variables occurring free in ω, θ does not occur in ω, and the corresponding uniqueness condition

$$(\gamma_1) \ldots (\gamma_n)(\exists\beta)(\alpha)(\alpha = \beta \leftrightarrow \omega),$$

with β distinct from $\alpha, \gamma_1, \ldots, \gamma_n$, is a thesis of T.

Our discussion has concerned the problem of adding a single definition to a theory T, but it is probably clear how to extend our remarks to the more typical problem of adding a whole sequence of definitions. This amounts to generating a corresponding sequence of theories; by adding the first definition to T, we produce a larger theory T_1; and, by adding the second definition to T_1, we produce a still more comprehensive theory T_2; and so on. To ensure eliminability and non-creativity we have only to take care that each definition is a formally correct definition relative to the theory to which it is added.

Finally, a few words about the notion of *definability*. It may easily be checked that, if from the non-logical vocabulary of a theory T we remove a constant θ, and if from the theses of T we remove all those containing θ, the result will again be a theory if the reduced vocabulary still contains at least one predicate of degree ≥ 1. Let us call this theory $T - \theta$. Thus the vocabulary of $T - \theta$ is the vocabulary of T with the exception of θ, and the theses of $T - \theta$ are all theses of T that do not contain θ. We shall say that the non-logical constant θ is *definable* in the theory T if and only if there is among the theses of T a formally correct definition of θ relative to the theory $T - \theta$. Thus, if a non-logical constant θ is definable in a theory T, the essential content of T can be presented in the reduced theory $T - \theta$, and then the notational advantages of T, if any, may be retrieved by adding an appropriate definition of θ.

A very useful device connected with this sort of definability is *Padoa's principle:* to show that a non-logical constant θ is not definable in a theory T, give two models of T that differ in what they assign to θ but are otherwise the same. We shall not take the trouble to prove that if two such models can be given, then no formally correct definition of θ relative to $T-\theta$ can be a thesis of T, but this will be intuitively evident to anyone who considers the four types of sentence we have specified.

EXERCISES

1. Show that the following definitions of 'consistent' and 'complete' are equivalent to those given in the text: a theory T is consistent if and only if there is a sentence of T that is not asserted by T; a theory T is complete if and only if the addition of any unasserted sentence of T to the assertions of T would render T inconsistent.
2. (a) Show that Λ is an independent set of sentences.
   (b) Give an example of a sentence ϕ such that $\{\phi\}$ is not independent.
3. Demonstrate the assertion made in the last sentence of paragraph 5, page 184.
4. (a) Describe a theory T for which Λ is a set of axioms.
   (b) Show that if a theory T is inconsistent, it is axiomatizable.
5. Prove:
   (a) theorems 11–19 of T_1.
   (b) theorems 11–20 of T_2.
   (c) theorems 9–12 of T_3.
   (d) theorems 12–28 of T_4.
6. Let T_0 be a subtheory of T_1 having as non-logical vocabulary the three predicates 'A^2', 'E^2', and 'I^2', and having as theses all sentences that are formulated in this vocabulary and are consequences of axioms 1–6 of T_1.
   (a) Show that axiom 7 of T_1 is a formally correct definition of 'O^2' relative to T_0.
   (b) Following the lines of the general arguments given on page 190, show in this particular case that the criteria of eliminability and non-creativity are satisfied. (Letting ϕ = axiom 7, take ψ = theorem 11 and μ = theorem 8).
7. Let T_{-1} be a subtheory of T_1 having as non-logical vocabulary the two predicates 'A^2', and 'I^2', and having as theses all sentences that are formulated in this vocabulary and are consequences of axioms 1, 3, and 5 of T_1 and the sentence

$$(x)(y)(z)((-Iyz \mathbin{\&} Axy) \rightarrow -Ixz).$$

Show how to generate T_1 from T_{-1} by adding a pair of formally correct definitions.

8. Show that

$$(x)(y)(w)(w = g^2xy \leftrightarrow x = y + w)$$

is a formally correct definition of the binary operation symbol 'g^2' relative to the theory T_3.

9. Relative to the theory T_4 construct formally correct definitions of set-theoretical difference and set-theoretical inclusion.

10. Consider the following six theories formalized in the language $\mathfrak{L}_I$. (Compare the theory T_2 of section 3.) In the first two cases the non-logical vocabulary consists of the single predicate 'A^2'; in the third and fourth cases it consists of the predicate 'B^3'; and in the remaining two cases it consists of both of these predicates. In all six cases the theses are all sentences of the theory that are consequences of the given axioms.

T_A Axioms: (*i*) $(x)Axx$
(*ii*) $(x)(y)((Axy \& Ayx) \rightarrow x = y)$
(*iii*) $(x)(y)(z)((Axy \& Ayz) \rightarrow Axz)$

T_A^* Axioms: (*i*), (*ii*), (*iii*), and
(*iv*) $(x)(y)(Axy \vee Ayx)$

T_B Axioms: (1) $(x)(y)(Bxyx \rightarrow x = y)$
(2) $(x)(y)(z)(u)((Bxyu \& Byzu) \rightarrow Bxyz)$
(3) $(x)(y)(z)(u)(((Bxyu \& Byuz) \& y \neq u) \rightarrow Bxyz)$

T_B^* Axioms: (1), (2), (3), and
(4) $(x)(y)(z)((Bxyz \vee Bxzy) \vee Bzxy)$

T_{AB} Axioms: (*i*), (*ii*), (*iii*), and the definitional axiom
(*v*) $(x)(y)(z)(Bxyz \leftrightarrow ((Axy \& Ayz) \vee (Azy \& Ayx)))$

T_{AB}^* Axioms: (*i*), (*ii*), (*iii*), (*iv*), and (*v*).

Let us say that a theory T' is an *extension* of a theory T if and only if the non-logical vocabulary of T' includes that of T and the set of theses of T' includes the set of theses of T. Let us say further that a theory T' is a *conservative extension* of a theory T if and only if T' is an extension of T and, in addition, each thesis of T' that is a sentence of T is a thesis of T.

Now show:

(*a*) T_{AB} is a conservative extension of T_A, and T_{AB}^+ is a conservative extension of T_A^+.

(*b*) Each axiom of T_B is a thesis of T_{AB}, and hence T_{AB} is an extension of T_B; also, T_{AB}^+ is an extension of T_B^+.

12

A BRIEF OUTLINE OF THE HISTORY OF LOGIC

1. *Ancient logic*
2. *Medieval logic*
3. *Modern logic*

In approaching the history of logic one must keep in mind that the term 'logic' and its cognates have been applied to many subjects other than the one presently under consideration, and, conversely, that the latter has been denoted by many terms other than 'logic'. Even if our competence permitted it, there would seem to be little point in undertaking a simultaneous history of all those topics in epistemology, metaphysics, psychology, sociology, and philology that at one time or another have been discussed under the heading 'logic'. The goal here is only to set forth the history of what *we* call 'logic'—roughly characterizable, perhaps, as the general theory of the consequence relation—by whatever names this subject may be known to other authors, past or present.

For the sake of clarity it should also be borne in mind that the proper business of a logician is the investigation and formulation of general principles concerning what follows from what; whether particular examples of his own reasoning are valid or not is essentially irrelevant. By the same token, correct reasoning, however praiseworthy it may be, does not of itself constitute a contribution to logic; men were giving valid arguments long before there was any such thing as the science of logic, just as stones were no doubt efficiently pried up long before anyone formulated the principle of the lever.

1. *Ancient logic.* If, with these points in mind, we now look to the origins of our science, we can say flatly that the history of logic begins with the Greek philosopher Aristotle (384–322 B.C.). Although it is almost a platitude among historians that great intellectual advances are never the work of only one person (in founding the science of geometry Euclid made use of the results of Eudoxus and others; in the case of mechanics Newton stood upon the shoulders of Descartes, Galileo, and Kepler; and so on), Aristotle, according to all available evidence, created the science of logic absolutely *ex nihilo.* With disarming straightforwardness he tells us this himself in a passage at the end of the *Sophistical Refutations,* and there is no reason to doubt the accuracy of his report. Many scholars, to be sure, have believed on *a priori* grounds that such an act of creation is impossible, and accordingly they have combed the writings of Aristotle's predecessors, especially those of Plato, in search of at least 'the germ' of Aristotelian logic. This quest has been almost entirely fruitless; because of confusion on the points discussed in our preceding two paragraphs, however, it has not always been recognized as such.

Aristotle's writings on logic are contained in a group of treatises that in later times came to be known collectively as the *Organon.* These are six in number: the *Categories, De Interpretatione, Prior Analytics, Posterior Analytics, Topics,* and *Sophistical Refutations.* (The titles are probably not due to Aristotle and give little indication of the content). In printed form they add up to a volume of several hundred pages, but the syllogistic, or theory of the syllogism, which is the real core of Aristotelian logic, will be found set forth in a very few pages at the beginning of the *Prior Analytics.* Most of the rest of the *Organon* is devoted to topics that lie outside the field of logic, though occasional passages do throw light on the terminology employed in the syllogistic or provide other useful background information.

Before going further we ought to mention parenthetically that in reading Aristotle one must always make allowance for the vicissitudes that the Aristotelian writings have suffered during the twenty-three centuries of their history. Passages have been garbled, the marginal notes of commentators have been inserted into the text, the order of books and chapters has been scrambled, whole sections have been lost, and spurious works have been added—all this in addition to the normal copyists' mistakes of omission, reduplication, and substitution. The logician who reads Aristotle will also have to adjust to the author's neglect of the use-mention distinction. For example, locutions of the form 'all *A* is *B*' and '*A* is included in *B*' are used interchangeably with corresponding locutions of the form '*B* is predicated of all *A*' and '*B* belongs to all *A*'; in fact, at one place the author says flatly that 'for one thing to be included as a whole in another, and for the other to be predicated of all of the one, are the same'. Thus in the *Categories* we find the statement

> Whenever one thing is predicated of another as of a subject, everything predicated of what is predicated will also be predicated of the subject; e.g., man is predicated of a particular man, and animal of man; thus animal will be predicated also of the particular man.

If we raise the question whether Aristotle is here talking about words or things or both, we are probably raising a question to which there is no answer; this does not imply, of course, that what he says has no content.

A syllogism, according to Aristotle, is a piece of discourse in which, certain things being posited, something else follows necessarily from their being so. This definition would lead one to suppose that Aristotle uses the term 'syllogism' in approximately the sense of 'valid argument', but in fact his usage is much more restricted. Near the beginning of the *Prior Analytics* he lists the kinds of sentence that may be components of a syllogism. Every premise or conclusion, we are told, is affirmative or negative according as it asserts or denies something of something. It is also universal, particular, or indefinite: a universal sentence states that something belongs to all or none of something else; a particular sentence states that something belongs to some or not to some or not to all of something else; while an indefinite sentence merely asserts, neither in general nor in particular, that something belongs or does not belong to something else, e.g., that pleasure is not good. In practice the indefinite sentences are ignored by Aristotle; the reason for this, according to the commentators, is that they are 'equivalent' to the corresponding particular sentences. In any case, the components of Aristotelian syllogisms are always sentences that are universal or particular and affirmative or negative; that is, using Aristotle's own examples, they are sentences like 'Every man is white', 'No man is white', 'Some man is white', and 'Not every man is white', later known respectively as sentences of *A*, *E*, *I*, or *O* form. Expressions like 'man' and 'white' are called 'terms'. The theory of the syllogism makes no provision for singular sentences, like 'Socrates is white', although sentences of this type have played a prominent role in accounts of so-called traditional logic.

Not every argument composed of *A*, *E*, *I*, or *O* sentences is a syllogism, it turns out, but only such as contain exactly two premises and a conclusion and involve at most three terms. Thus the two premises always have at least one term in common; such a term is called a *middle term.* The predicate of the conclusion is the *major term,* and the subject of the conclusion is the *minor term.*

In the treatise *De Interpretatione* Aristotle mentions some of the logical relationships that hold among *A*, *E*, *I*, and *O* sentences having the same subject and predicate terms. The *A* and the *O* are *contradictories* of one another, as are the *E* and the *I*; of every pair of contradictories, he says, exactly one is true. The *A* and the *E* are called *contraries;* contraries cannot both be true, but they can both be false. These relationships and others were later represented schematically in the Square of Opposition, a figure that is found in nearly every traditional logic text and first appeared in the commentary of Apuleius of Madauros (second century A.D.) on *De Interpretatione.*

Aristotle prefaces his deductive exposition of the theory by stating the so-called laws of conversion, which he later uses in 'reducing' one kind of syllogism to another. He says that the universal negative sentence converts into a universal negative; for example, if no pleasure is good, then no good will be pleasure. The universal and particular affirmative sentences convert into particular affirmatives; for example, if every pleasure is good, or if some pleasure is good, then some good is pleasure. The particular negative does not convert; it is not the case that if some animal is not a man, then some man is not an animal. Aristotle formulates these laws with the help of variables:

If *A* belongs to no *B*, then *B* will not belong to any *A*.
If *A* belongs to all *B*, then *B* will belong to some *A*.
If *A* belongs to some *B*, then *B* will belong to some *A*.

This is the first clear use of variables in the history of science.

The exposition of the theory proceeds by listing the valid types (or 'moods') of syllogism, indicating how some of these may be derived from ('reduced to') others, and by rejecting invalid moods on the basis of counterexamples. Aristotle's own statements of the valid moods, literally translated, are presented below. Important words that do not occur in his typically condensed statements, but are seemingly required by grammar and sense, have been enclosed within brackets. The names of the moods are a medieval contribution; as is standard, the citations are to the page, column, and line of the Berlin edition.

Barbara. For if *A* [is predicated] of all *B* and *B* of all *C*, it is necessary for *A* to be predicated of all *C*. (25 b 37)

Celarent. Likewise, if *A* [is predicated] of no *B* but *B* of all *C*, [it is necessary] that *A* will belong to no *C*. (25 b 40)

Darii. Let *A* belong to all *B*, and *B* to some *C*. Then if 'predicated of all' means what was said at the beginning, it is necessary for *A* to belong to some *C*. (26 a 23)

Ferio. And if *A* belongs to no *B*, but *B* to some *C*, it is necessary that *A* does not belong to some *C*. (26 a 25)

Cesare. Let *M* be predicated of no *N* but of all *O*. Then, since the negative converts, *N* will belong to no *M*. But *M* was assumed [to belong] to all *O*; so that *N* [will belong] to no *O*; for this has been shown previously. (27 a 5)

Camestres. Again, if *M* [belongs] to all *N* but to no *O*, neither will *N* belong to any *O*. For if *M* [belongs] to no *O*, then *O* [belongs] to no *M*; but *M* belonged to all *N*; therefore, *O* will belong to no *N*; for the first figure has been produced again. And since the negative converts, *N* will belong to no *O*. (27 a 9)

Festino. For if *M* belongs to no *N* but to some *O*, it is necessary for *N* not to belong to some *O*. For since the negative converts, *N* will belong to no *M*. But *M* was assumed to belong to some *O*, so that *N* will not belong to some *O*; for a syllogism in the first figure is obtained. (27 a 32)

Baroco. Again, if *M* belongs to all *N* but not to some *O*, it is necessary for *N* not to belong to some *O*. For if it belongs to all, and if *M* is also predicated of all *N*, it is necessary for *M* to belong to all *O*. But it was assumed not to belong to some. (27 a 36)

Darapti. Whenever both *P* and *R* belong to all *S*, [it is true] that *P* will of necessity belong to some *R*. For since the affirmative converts, *S* will belong to some *R*, so that since *P* [belongs] to all *S*, and *S* to some *R*, it is necessary for *P* to belong to some *R*. For a syllogism in the first figure is produced. It is also possible to make the proof *per impossibile* and by *ekthesis*. For if both belong to all *S*, then if one of the *S*'s, e.g., *N*, is taken, both *P* and *R* will belong to this, so that *P* will belong to some *R*. (28 a 17)

Felapton. And if *R* belongs to all *S*, but *P* to none, there will be a syllogism that of necessity *P* will not belong to some *R*. The same sort of proof [will work] by converting the *RS* premise. It might also be shown *per impossibile,* as in the previous cases. (28 a 26)

Disamis. For if *R* [belongs] to all *S* and *P* to some, it is necessary for *P* to belong to some *R*. For since the affirmative converts, *S* will belong to some *P*, so that since *R* [belongs] to all *S*, and *S* to some *P*, *R* will belong to some *P*. So that *P* [will belong] to some *R*. (28 b 7)

Datisi. Again, if *R* belongs to some *S* and *P* to all, it is necessary for *P* to belong to some *R*. For the same method of demonstration [will work]. And it is also possible to demonstrate it both *per impossibile* and by *ekthesis,* just as in the previous cases. (28 b 11)

Bocardo. For if *R* [belongs] to all *S* and *P* does not belong to some, it is necessary for *P* not to belong to some *R*. For if [it belongs] to all, and *R* [belongs] to all *S, P* will also belong to all *S*. But it did not so belong. It is also proved without reduction, if one of the *S*'s to which *P* does not belong is taken. (28 b 17)

Ferison. For if *P* [belongs] to no *S*, but *R* belongs to some *S*, *P* will not belong to some *R*. For the first figure will be produced again when the *RS* premise is converted. (28 b 33)

Fesapo and *Fresison.* ... such as, if *A* [belongs] to all *B* or to some, but *B* to no *C*. For when the premises are converted it is necessary for *C* not to belong to some *A*. Likewise in the other figures: a syllogism always results from conversion of the premises. (29 a 23)

Aristotle groups the moods of the syllogism into three so-called figures. In order to prove by a syllogism that *A* belongs or does not belong to *B*, he says, it is necessary to take something common in relation to both, and this may be done in three ways: by predicating either *A* of *C* and *C* of *B*, or *C* of both, or both of *C*. 'These are the aforementioned figures, and it is clear that every syllogism must necessarily be in one of these figures'. Thus the valid moods Barbara, Celarent, Darii, and Ferio belong to the first figure; Cesare, Camestres, Festino, and Baroco to the second; and Darapti, Felapton, Disamis, Datisi, Bocardo, and Ferison to the third. In later times a fourth figure was added, corresponding to the possibility of proving *A* of *B* by predicating *C* of *A* and *B* of *C*. Nobody knows whether Aristotle omitted this possibility as a result of some theoretical consideration, or simply by oversight.

To the modern logician the most interesting feature of the theory of the syllogism is its development as an axiomatic system. Aristotle was aware that this can be done in more than one way. First he selects as axioms the valid moods of the first figure and proves the others by reducing them to the axioms. His reductions are direct or indirect. A direct reduction is accomplished by converting one or more premises of the syllogism to be proved, reversing their order if necessary, and then deriving the desired conclusion by using the syllogism to which reduction is made. As may be seen from the translation given above, Aristotle uses this method to reduce Cesare and Camestres to Celarent; Festino, Felapton, and Ferison to Ferio; and Darapti, Disamis, and Datisi to Darii. The reduction of Baroco and Bocardo to Barbara, on the other hand, is indirect. That the conclusion of a given syllogism follows from its premises is established by assuming one of the premises and the contradictory of the conclusion, and then, using the syllogism to which reduction is being made, deriving the contradictory of the other premise. Thus, the argument is in accord with the theorem

$$((P \& Q) \rightarrow R) \rightarrow ((P \& -R) \rightarrow -Q),$$

even though our author nowhere explicitly formulates any such principle.

Having shown how to reduce all the valid moods to those of the first figure, Aristotle next asserts that Barbara and Celarent alone would suffice as axioms for the theory. He establishes this by reducing Darii and Ferio to moods of the second figure, and then by showing that all moods of the second figure may be reduced to Barbara and Celarent of the first. (Cesare, Camestres, and Baroco have already been so reduced, and Aristotle explains that Festino may be reduced indirectly to Celarent as well as directly to Ferio). To top off the performance, he correctly observes that, in fact, the valid moods of any of the three figures could serve equally well as axioms. For, as he does not bother to show, Barbara may be reduced indirectly to Baroco or Bocardo, and Celarent to Festino or Disamis. Thus, not only did Aristotle introduce variables and use them to formulate, for the first time, a number of formally valid laws of logic, but by means of the first axiom system in history he managed to demonstrate some of the interrelationships among those laws. Also worthy of special note is his implied awareness that to some extent the choice of axioms is arbitrary;

what is an axiom in one formulation of a given theory may be a derived theorem in another.

The fourteen valid moods are arrived at by the use of counter-examples to eliminate all other possibilities. Since the passages containing these eliminations are rather cryptic, it may be worth while to explain a typical instance.

> But if *M* is predicated of all *N* and *O*, there will not be a syllogism. Terms for the affirmative relation are substance, animal, man; for the negative relation substance, animal, number.

This is Aristotle's way of showing that in the second figure none of the four moods beginning 'If *M* is predicated of all *N* and *O*, then . . .' is valid. Expanded, the argument would run as follows. If there were a valid mood with the given premises and a negative conclusion, then every assignment of terms to variables that made the premises true would also make this conclusion true, and hence would make the formula '*N* is predicated of all *O*' false. (For the universal affirmative is incompatible with both the universal and particular negatives). But assigning the terms substance, animal, and man to '*M*', '*N*', and '*O*', respectively, makes the premises true and the universal affirmative formula true; therefore, there is no valid mood with the given premises and a negative conclusion. Similarly, if there were a valid mood with the given premises and an affirmative conclusion, any assignment that made the premises true would have to make this conclusion true and, consequently, would make the universal negative formula '*N* is predicated of no *O*' false. But assigning substance, animal, and number to '*M*', '*N*', and '*O*' makes the premises and this formula true. Therefore, there is no valid mood with the given premises and an affirmative conclusion. We see that Aristotle is here using the same idea that motivates our definition of 'consequence' (page 63). To show that a mood is invalid he produces an interpretation under which its premises are true and its conclusion false. By ingenious assigning of terms, he sometimes contrives to eliminate as many as eight moods with a single interpretation.

Limitation of space prevents us from attempting an exposition of Aristotle's contributions to modal logic, i.e., to the theory of the modal operators 'necessarily' and 'possibly'. He had much to say on this subject, but most of it seems garbled and confused. Unfortunately, the reader is always left wondering what proportion of the difficulty is due to the nature of the subject, what proportion to Aristotelian confusion, what to corruption of the text, and, finally, what to his own thick-headedness.

We know very little about the history of logic in the Peripatetic school after Aristotle. Theophrastus (ca. 372–288 B.C.), the next head of the school, apparently devoted himself almost exclusively to the development and correction of his master's discoveries. He is said to have added five valid moods to the first figure of the syllogism; this, however, revealed no defect in Aristotle's analysis but only an ambiguity in the term 'first figure'. Instead of the pattern $AB - BC - AC$, which is exhibited by all of Aristotle's first-figure moods, the five new moods have the pattern $AB - BC - CA$; as, for example, in

> If A is predicated of all B and B of all C, then it is necessary for C to be predicated of some A.

Thus, when their premises are reversed, they are seen to be just the five valid moods of the fourth figure. Theophrastus is also mentioned as having achieved certain clarifications in Aristotle's modal logic and as having concerned himself with the so-called hypothetical syllogisms. The latter are arguments of such forms as

> If A then B and if B then C; therefore, if A then C.

or

> If A then B and if not A then C; therefore, if not C then B.

While the Peripatetics were occupied with preserving their heritage from Aristotle, another group of philosophers, the Stoics and Megarians, were developing a radically different approach to formal logic. They were, in effect, inventing the sentential calculus. Unfortunately, all of the writings on logic of these authors have been lost, and consequently we have to reconstruct their doctrines from fragments found in the works of persons who were writing centuries later. For the obvious reasons, it is too much to expect that the composite picture thus obtained will be entirely satisfactory; in fact, the wonder is that it is consistent at all.

The Megarian school was founded by Euclid (not to be confused with the geometer), a follower of Socrates. His students included Eubulides, to whom the antinomy of The Liar is ascribed, and Thrasymachus of Corinth, who was the teacher of Stilpo. Stilpo, in turn, was the teacher of Zeno (ca. 336–264 B.C.), the founder of Stoicism. Very little is known about any of these men, but where the facts leave off legend begins. Concerning Zeno, for example, we are told that he was not a Greek but was born in Cyprus, came later to Athens, aroused local ire by proposing to reform the Greek language before he had learned to speak it, and, after a long philosophical career, died at the age of 98 by holding his breath. His successors in charge of the Stoic school were Cleanthes (described as a poverty-stricken prize fighter who came to Athens and entered Zeno's school, became its head, transmitted Zeno's doctrines without change, and eventually starved himself to death at the age of 99) and Chrysippus (ca. 280–205 B.C.). Next to Aristotle, Chrysippus was the most productive logician of antiquity. According to some old sayings, 'If there is any logic in Heaven, it is that of Chrysippus', 'If there had been no Chrysippus, there would have been no

Stoa'; and Chrysippus himself is quoted as saying to Cleanthes, 'Just send me the theorems; I'll find the proofs for myself'. Another important branch of the Megarian school contained the two logicians Diodorus Cronus (d. 307 B.C.) and his student Philo. Diodorus contributed definitions of necessity and possibility in terms of notions of 'always true' and 'sometimes true'; Philo, so far as we know, is the inventor of material implication.

The Stoics, unlike Aristotle, were quite clear about use and mention. They had a semantical theory somewhat similar to that of Frege, involving a distinction between the sign, its sense, and its denotation. The sense 'is what Greeks, but not Barbarians, are able to grasp when they hear Greek words spoken'. The sense of a declarative sentence is a proposition; only propositions can be true or false, and therefore they constitute the subject matter of logic. Much attention was devoted by the Stoics and Megarians to the sense of the connectives 'if ... then', 'and', and 'or'. In particular, the controversy over the proper interpretation of conditionals was so intense that, according to a long misunderstood fragment from Callimachus, 'even the crows on the rooftops are cawing over the question as to which conditionals are true'. In a very interesting passage, Sextus Empiricus (third century), our principal source for Stoic logic, describes the four main interpretations that were considered. He arranges them from weakest to strongest, in each case giving an example that holds in all preceding senses but fails in the next sense under consideration. (Numbers have been inserted to mark off the different views more clearly.)

> (1) For Philo says that a true conditional is one that does not have a true antecedent and a false consequent; e.g., when it is day and I am conversing, 'If it is day, then I am conversing'; (2) but Diodorus defines it as one that neither is nor ever was capable of having a true antecedent and a false consequent. According to him, the conditional just mentioned seems to be false, since when it is day and I have become silent, it will have a true antecedent and a false consequent; but the following conditional seems true: 'If atomic elements of things do not exist, then atomic elements of things do exist', since it will always have the false antecedent, 'Atomic elements of things do not exist', and the true consequent, 'Atomic elements of things do exist'. (3) And those who introduce 'connection' or 'coherence' say that a conditional holds whenever the denial of its consequent is incompatible with its antecedent; so that, according to them, the above-mentioned conditionals do not hold, but the following is true: 'If it is day, then it is day'. (4) And those who judge by 'suggestion' declare that a conditional is true if its consequent is in effect included in its antecedent. According to these, 'If it is day, then it is day', and every repeated conditional will probably be false, for it is impossible for a thing to be included in itself.
>
> (Sextus Empiricus, *Outlines of Pyrrhonism,* II, 110)

Philo's definition of material implication occurs frequently in the fragments, usually in a stylized form reminiscent of a truth-table analysis:

Since, then, there are four possible combinations of the parts of a conditional—true antecedent and true consequent, false antecedent and false consequent, false and true, or conversely true and false—they say that in the first three cases the conditional is true (i.e., if the antecedent is true and the consequent is true, it is true; if false and false, it again is true; likewise for false and true); but in one case only is it false, namely, whenever the antecedent is true and the consequent is false.

(Sextus Empiricus, *Against the Mathematicians,* VIII, 247)

The connectives 'and' and 'or' were also given both truth-functional and modal interpretations, and in the case of 'or' an inclusive and an exclusive sense were distinguished. It was realized that if we keep to truth-functional senses, the connective 'if . . . then' may be defined in terms of 'not' and 'and', and (perhaps) that 'or' may be defined in terms of 'if . . . then' and 'not'. In fact Chrysippus recommended that for clarity the material conditional 'If anyone is born under the Dog Star, then he will not be drowned in the sea' be expressed as a negated conjunction, 'Not both: someone is born under the Dog Star and will be drowned in the sea'. (Note, however, that these sentences are properly general, not molecular.)

An argument, according to the Stoics, is 'a system composed of premises and a conclusion'. The standard example was an instance of *modus ponens:*

If it is day, then it is light;
It is day;
Therefore, it is light.

An argument is defined as sound if its corresponding conditional is true; evidently one of the stronger interpretations of the conditional was employed here. The term 'sound' was also applied to argument schemata, like

If the first, then the second;
The first;
Therefore, the second.

An argument schema is sound if all its instances are sound. Clearly the Stoics were using 'the first', 'the second', etc., as variables for which sentences could be substituted; this is in sharp contrast to the Aristotelian variables, the substituends for which were general terms like 'man' and 'animal'.

The Stoics, like Aristotle, attempted to arrange all sound arguments in a sort of deductive system. They seem to have considered only such arguments as are formally sound in the sense of the sentential calculus. Arguments of five types were taken as basic, and all others were declared reducible to chains of these. The five basic types are instances of the following schemata:

I. If the first, then the second;
The first;
Therefore, the second.

II. If the first, then the second;
Not the second;
Therefore, not the first.
III. Not both the first and the second;
The first;
Therefore, not the second.
IV. The first or the second;
The first;
Therefore, not the second.
V. The first or the second;
Not the first;
Therefore, the second.

Proofs in the system were done schematically, by means of four rules. Unfortunately, our knowledge of these rules and their application is incomplete. On the basis of some examples given by Sextus, however, plausible conjectures have been made. Consider, he says, the following argument schema:

(1) If both the first and the second, then the third;
(2) Not the third;
(3) The first;
Therefore, not the second.

An argument of this type is compounded of basic arguments of types II and III. From (1) and (2), by II, we get

(4) Not both the first and the second.

This, according to a so-called 'dialectical rule', can now be added to the premises. Then, from (3) and (4), by III, we obtain the conclusion. The dialectical rule is: 'if we have premises that yield a conclusion, then we have in effect also the conclusion among the premises, even if it is not explicitly stated.' Thus, it is clear that the Stoics were trying to find a manageable set of inference rules by means of which one could derive, by an exactly specified procedure, the tautological consequences of given premises. From remarks quoted by Cicero it appears that they believed their five basic rules to be sufficient; our imperfect knowledge of the details prevents us from evaluating this claim.

The fragments contain a few more schemata that are asserted to be provable.

Either the first or the second or the third.
Not the first.
Not the second.
Therefore, the third.

Chrysippus suggested that even dogs seem to utilize arguments of this pattern. He claimed to have noticed that when a dog is chasing an animal and comes to a point where the path he is following divides into three, if he sniffs first at the two paths which the animal did not take, he will dash off down the other way without pausing to test it. In effect, according to Chrysippus, the dog argues as follows:

Either it went this way or that way or the other way.
It didn't go this way.
It didn't go that way.
Therefore, it went the other way.

The argument is said to involve repeated application of schema V.

We are indebted to Origen for another amusing example of Stoic argumentation:

If you know you are dead, you are dead.
If you know you are dead, you are not dead.
Therefore, you do not know that you are dead.

(And, of course, since the premises are analytic, so is the conclusion.) Origen also gives the schema,

If the first, then the second.
If the first, then not the second.
Therefore, not the first.

but he does not tell us how it was analyzed into the five basic types.

Finally, we may mention the Stoic interest in paradoxes and antinomies. The most famous of these was The Liar; in one version it runs: the man who says 'I am lying' is both lying and telling the truth. This important antinomy was taken very seriously in antiquity, as well as in medieval and modern times; Chrysippus wrote whole books on it, and there is even the epitaph of the logician Philetas of Cos (as translated by St.-George Stock):

Philetas of Cos am I,
'Twas The Liar who made me die,
And the bad nights caused thereby.

For more than 1000 years after Chrysippus there was no one, so far as is known, who contributed anything very original to the science of logic. Such authors as are worthy of mention are important only for their service in preserving the ancient doctrine and making possible its eventual transmission to the Middle Ages (and eventually to ourselves). Thus the great orator Cicero (106–43 B.C.) gives some bits of information about Stoic logic and is responsible for translating much of the Greek logical terminology into Latin. From the second century A.D. we have a pair of 'introductions to logic', purportedly written by Apuleius of Madauros, mentioned above in connection with the Square of Opposition, and the

Greek physician Galen (131–201). These books and others like them seem to have played an essential role in preventing the corruption or disappearance of the ancient discoveries. They also show, however, that by the middle of the second century the inevitable confusion of Stoic and Aristotelian elements was already well advanced. At the beginning of the third century, the Aristotelian commentator Alexander wrote a very useful exegesis of Aristotle's logical works, incorporating a certain amount of information about the Stoics. Later we have Sextus Empiricus and Diogenes Laertius, who are our best sources for Stoic logic. A Latin translation of the former's *Outlines of Pyrrhonism* may have been available as early as the twelfth century and thus may have played a role in the development of medieval logic. At the beginning of the fifth century we find Boethius (470–524) and Martianus Capella, both of whom endeavored to compile a picture of the logical tradition as it existed in their day. Boethius made Latin translations of Aristotle's *Categories* and *De Interpretatione* and also wrote commentaries on these works and on the *Introduction* (to Aristotle's *Categories*) written in the third century by the Greek commentator Porphyry; in addition he composed treatises on the categorical and hypothetical syllogism. His work shows that by and large he knew what he was about, and until the middle of the twelfth century it was the principal source of information about ancient logic, but that is the most that can be said for it. As a logician, Martianus Capella is even less impressive, but his role as a transmitter of the tradition makes him worthy of mention.

2. *Medieval logic.* As I. M. Bocheński has pointed out, the history of logic does not consist of a gradual development leading from Aristotle down to modern times. Instead, there are three high points, each of relatively short duration, which are separated by long periods of decline. The first of these peaks occurred in the third and fourth centuries B.C., the second from the twelfth to the fourteenth centuries, and the third began in the late nineteenth century and, according to optimists, is at present in full swing. This is a rough generalization, of course, and there are a few important logicians —Leibniz is a prime example—who do not fall within any of the three periods, but in the main it is true.

The medieval contribution to logic, important as it was, lies more in the area we now call 'philosophy of logic' than it does in logic proper. To be sure, one cannot say this with complete confidence, for at present even less is known about medieval logic than about the logic of the ancients. Large numbers of manuscripts have never been read by competent historians, let alone edited, and they may contain important innovations. But so far as now appears, at any rate, the Middle Ages brought no new axiom systems, no increase in rigor as compared with Chrysippus or even Aristotle, and in general no continued progress at the level set by the best of the ancients. What it did contribute was a searching investigation of the semantics and logic of the Latin language, and a great deal of shrewd philosophizing about numerous intuitive issues that underlie any formal development of the subject. As an example of this, one might mention the thorough discussion of the question whether every sentence follows from a contradiction; several writers make the useful point that in order to get rid of this slightly odd result we must also give up various inference patterns that otherwise are completely unobjectionable (see page 19, exercise 2).

Before we consider individual authors, it should be emphasized that the most important single factor determining the nature of scholastic logic at the various periods of its history was the availability of source material inherited from the ancients. Until the middle of the twelfth century the only works to which there was general access were the *Categories* and *De Interpretatione* of Aristotle, Porphyry's *Introduction,* and various derivative works of Apuleius, Boethius, and Martianus Capella. Since the mood of the times placed heavy emphasis upon tradition, this paucity of source material is reflected in a corresponding restriction of the range and depth of discussion. By the latter half of the twelfth century, however, the revival of learning had progressed to the point where scholars were motivated to seek out as much as they could of antiquity's heritage, including the remainder of Aristotle's *Organon.* From that time onward the contributions of the schoolmen became much more numerous and sophisticated.

The first major figure in the history of medieval logic is Peter Abelard (1079–1142). It is true that Alcuin, who taught at York toward the end of the eighth century and later became head of the school established by Charlemagne, wrote a work entitled *Dialectica,* but this book contains little else than a discussion of Aristotle's categories. And presumably in the ninth and tenth centuries there were at least a few other such works. But not until Abelard and his school do we find full and relatively lucid discussion of large numbers of issues related to logic. A surprisingly great proportion of the topics and methods that appear throughout medieval logic get their start in the writings of Abelard. Thus, although he did not originate the great controversy about the existence of universals, he gave it its first strong impulse. His view was somewhere between realism (Platonism) and nominalism. 'Individual men, distinct from one another', he asserted, 'agree in that they are men; I do not say *in man* (*in homine*), since nothing is a man (*sit homo*) unless it is individual, but *in being a man.* For being a man is not man (*non est homo*) or anything else'. His work *Sic et Non* set the medieval pattern of presenting all philosophical discussion under the heading of *quaestiones*; a *quaestio* is posed, the arguments pro and con are systematically set forth, and then the *solutio* is given and applied to the previously stated arguments. The method is rigid and stylized, but it tends to make quite clear the structure of what an author has to say. Another of Abelard's innovations was his distinction between conditionals (*consequentiae*) that are true by virtue of their form (*ex complexione*) and those that are true by virtue of the facts (*ex rerum natura*). True conditionals of the latter sort, and also the arguments corresponding to them, he regarded as somehow imperfect. In a perfect conditional, he says, the sense of the consequent must be contained in that of the antecedent.

Abelard devoted much attention to the verb 'is', arguing that the content of any categorical sentence can be expressed by a sentence of the form '*A* is *B*' (*A est B*). Even 'Socrates exists' (*Socrates est*) can be represented by 'Socrates is an existent thing' (*Socrates est ens*). Perhaps this points the way to the possibility of reducing the number of predicates in our language to one—the '$\in$' of set theory—and of representing existence as membership in the universal set. Abelard also devoted a great deal of space to the modalities, raising issues that are still under discussion today.

In the work of Abelard we find no sign of direct acquaintance with Aristotelian writings other than the *Categories* and *De Interpretatione;* his meager account of the syllogistic is obviously drawn from Boethius. After the remainder of Aristotle's *Organon* became generally available, numerous *summulae* (little summaries) of logic appeared. Of these the oldest that has been printed is the work of William of Shyreswood (d. 1249). It contains, among a miscellany of other interesting items, two mnemonic 'poems' worth reproducing here. The first is the famous

> Barbara Celarent Darii Ferio Baralipton
> Celantes Dabitis Fapesmo Frisesomorum;
> Cesare Campestres Festino Baroco; Darapti
> Felapton Disamis Datisi Bocardo Ferison.

These verses list the valid syllogistic moods in the three figures (Theophrastus' five additional moods have been added to those of the first figure). In the names of the moods, most of the letters are significant. The first three vowels characterize the components of the syllogism; thus Barbara consists of three *A*-sentences with terms disposed according to the pattern of the first figure. The consonants give instructions for reducing the given mood to the first four: the initial consonant indicates the mood to which reduction is to be made (thus Baralipton, Baroco, and Bocardo reduce to Barbara); an occurrence of 's' means that the sentence denoted by the preceding vowel is to be converted simply; similarly for an occurrence of 'p', except that the conversion is not simple; an occurrence of 'm' indicates that the premises are to be interchanged; and a 'c' tells us to use indirect reduction. The reader may wish to see for himself whether these instructions work.

The other 'poem' gives, in effect, the content of our rule Q:

> *All, none-not,* and *not-some-not* are equivalent,
> As are *none, not-some* and *all-not*;
> *Some, not-none,* and *not-all-not* go together, and
> So do *some-not, not-none-not,* and *not-all.*

It reminds us, by the way, that our quantifiers *all* and *some* could just as well have been defined in terms of a quantifier with the sense of *none.*

Peter of Spain (ca. 1210–77), who probably studied under William of Shyreswood when the latter was teaching at Paris, and who later became Pope John XXI, wrote the only other book of *summulae* that is accessible in a modern edition. In its time it was regarded as something of a classic, and it remained in use until the seventeenth century. Its content was similar to that of the manual written by William, except for having more and better mnemonic verses; this, together with its author's high position, may explain its greater popularity. It contains sections on propositions, the five predicables of Porphyry (definition, genus, species, property, and accident), the categories, the syllogism, topical rules for argumentation, and fallacies; and, in addition, there is a group of tracts called *On the Properties of Terms.*

The doctrine of the properties of terms appears throughout later medieval logic and is often considered to be its most original contribution. Unfortunately, however, different authors give different accounts and we still await a really clear exegesis of any of them. The properties most commonly mentioned are *significatio, suppositio, copulatio,* and *appellatio;* they are said to characterize different aspects of the functioning of terms in Latin sentences as actually used. In this connection the word 'term' covers general nouns (e.g., 'man'), verbs (e.g., 'is' or 'runs'), and adjectives (e.g., 'white'). Every term has *significatio,* which seems in effect to be what might be called its 'dictionary sense' and, according to the realists, is always a form. But as used in a sentence a term may not stand for its *significatum.* If it has *suppositio materialis* (as the term *Homo* in *Homo est disyllabum*) it stands for itself; otherwise it has *suppositio formalis.* This latter may be *simplex* (as in *Homo est species*), where the term is used to refer to its *significatum,* or *personalis* (as in *Homo currit* or *Omnis homo est animal*), where it refers to one or more individuals falling under the form that is the *significatum.* The classification of types of *suppositio* is elaborated much further, always on the basis of distinctions that look interesting but are hard to elucidate. *Copulatio* (binding), in the original account given by Abelard, is the property of verbs by virtue of which they can join subject and predicate to form a categorical sentence; other authors define it quite differently. The *appellatio* of a term is said to be its reference to things presently existing. In general, until more light is thrown upon this subject one can only surmise that the properties of terms, if correctly understood, will turn out to be a collection of semantical concepts useful in explaining various logical puzzles that arise in the employment of the natural language (such as, for example, why 'Socrates used to be a boy' is not equivalent to 'Some boy used to be Socrates').

The principal logicians of the fourteenth century are William of Ockham (ca. 1295–1349), Jean Buridan (d. soon after 1358), Albert of Saxony (ca. 1316–90), and an unknown author whom we call Pseudo Scotus because his writings were long ascribed to Duns Scotus. Ockham's Razor (the proposition that 'entities should not be multiplied beyond necessity') and Buridan's Ass (an unfortunate donkey who starved to death because he could not choose between two equidistant piles of hay) will be well known to readers familiar with the history of philosophy. But in the history of logic the importance of Ockham and Buridan, as well as that of the other two, rests primarily upon their development of the theory of *consequentiae.*

The term *consequentia,* as defined by Pseudo Scotus, means 'a hypothetical proposition composed of an antecedent and a consequent, joined by a conditional conjunction', and it is clear that by 'conditional conjunction' he refers not only to 'if . . . then' but also to 'therefore'. Thus, examples of *consequentiae* include

> Every man is an animal; therefore, every animal is a man.
> Socrates exists and Socrates does not exist; therefore, Socrates does not exist.

For the soundness of a *consequentia* some such condition as the following is usually given: a *consequentia* is sound if and only if it is not possible for the antecedent to be true and the consequent false. This is the leading idea, though sometimes minor amendments were made in order to take care of certain paradoxes to which the stated condition seemed to give rise. Despite the generality of this conception of soundness, however, in practice only formally sound *consequentiae* were considered, and of these only such as in modern times would be counted part of the sentential calculus (sometimes supplemented by the modal operators 'necessarily' and 'possibly').

In their surveys of sound *consequentiae* the medieval authors used metalinguistic descriptions instead of schemata containing variables. Thus, instead of such formulas as

P; therefore, P or Q
Q; therefore, P or Q

and

P and Q; therefore, P
P and Q; therefore, Q

we find

> There is a sound *consequentia* from either part of an affirmative disjunction to the affirmative disjunction of which it is a part.

and

> Either part of a conjunction follows from the conjunction of which it is a part.

E. A. Moody has assembled a large number of such characterizations, found principally in the writings of the four authors mentioned above, and has set them forth systematically by means of formulas in modern notation. For the same purpose we may use SC formulas of our formalized language $\mathfrak{L}$, with the understanding that these serve only to indicate the existence of the corresponding metatheoretic statements. Corresponding to the two statements quoted above, therefore, we would have the SC theorems 25, 26, 47, and 48 of Chapter 6.

Other SC theorems for which, in a similar way, there are corresponding descriptions of families of valid *consequentiae* are numbers 1, 2, 3, 11, 12, 15, 42, 43, 45, 46, 49, 50, 66, and 73, to which may be added the theorems resulting from application of the principle of importation to numbers 1, 9, and 15. We also find analogues of the following:

$$(P \to Q) \to (-(P \to R) \to -(Q \to R))$$
$$(Q \to R) \to (-(P \to R) \to -(P \to Q))$$
$$(P \,\&\, -Q) \to -(P \to Q)$$
$$P \to (Q \vee -Q)$$
$$(P \vee (Q \,\&\, -Q)) \to P$$
$$-P \to -(P \,\&\, Q)$$
$$-Q \to -(P \,\&\, Q)$$
$$(P \,\&\, ((P \to Q) \,\&\, (Q \to R))) \to R$$
$$(P \to Q) \to ((Q \to -R) \to (P \to -R))$$
$$(P \to Q) \to ((P \,\&\, R) \to (Q \,\&\, R))$$
$$(P \to Q) \to (((P \,\&\, Q) \to R) \to (P \to R))$$
$$(P \to Q) \to (((Q \,\&\, R) \to S) \to ((P \,\&\, R) \to S))$$
$$((P \,\&\, Q) \to R) \to ((P \,\&\, -R) \to -Q)$$
$$((P \,\&\, Q) \to R) \to ((Q \,\&\, -R) \to -P)$$
$$((P \,\&\, Q) \to R) \to (-R \to (-P \vee -Q))$$

In addition to all these, several sound *consequentiae* involving modal operators were mentioned. To give just one example:

> For the possibility of a disjunction it suffices that either part is possible.

Before leaving the medieval period we must set forth a rather remarkable objection raised by Pseudo Scotus to the standard characterization of a sound *consequentia* as one in which it is impossible for the antecedent to be true and the consequent false. After pointing out that on this basis any *consequentia* with a necessary consequent will be sound, he offers to produce an example of an unsound *consequentia* in which both the antecedent and the consequent are necessary. It turns out to be:

God exists; therefore, this *consequentia* is not sound.

The *consequentia* is surely unsound, he says, since otherwise we would have a sound *consequentia* with a true antecedent and a false consequent. And since we have established this unsoundness using only the necessary truth that God exists, the unsoundness is necessary. Thus the *consequentia,* though unsound, has a necessary consequent. We leave the reader to explore this argument for himself, noting only that the premise 'God exists' could equally well have been replaced by '$2 + 2 = 4$' or any other necessary truth.

3. *Modern logic.* The Renaissance, with its reaction against medieval scholasticism, marks the beginning of another long period of relative inactivity in the history of logic. To the humanists, rediscovering the beauties of classical Greek and Latin literature, the writings of the logicians seemed not only dull and trivial in content but also barbarous in style. These feelings were shared by men of scientific bent, who were finding in addition that for their purposes the Aristotelian syllogistic and all its appurtenances were actually worse than useless. Under such circumstances it is hardly surprising that logic no longer attracted the more talented minds and that as a result it sank gradually into a state of neglect. Not until the appearance, more than four hundred years later, of Boole and De Morgan and Frege, did it recover from the effects of this setback and experience a renaissance of its own. With the single exception of Leibniz, all the logicians of this period must be classified as minor.

The first of these is Peter Ramus (1515–72), who wrote a number of treatises on logic and was known primarily for his anti-Aristotelianism. The great popularity and influence of his work seem hardly justified; perhaps they are due to the fact that he was murdered in the massacre of St. Bartholomew's Day and thereafter was regarded by Protestants as a martyr. In any case, he did perform the service of questioning whether Aristotle had not neglected such syllogisms as 'Octavius is the heir of Caesar; I am Octavius; therefore, I am the heir of Caesar'. Defenders of Aristotle were reduced to the desperate state of arguing that this sort of syllogism should be recast as 'Whatever is Octavius is the heir of Caesar; whatever is I is Octavius; therefore, whatever is I is the heir of Caesar.'

The philosopher Thomas Hobbes (1588–1679) deserves a brief mention for his original and emphatic statement of the view that necessary truths are true simply by virtue of the ways in which their component terms are used. ''Man is a living creature' is true', he says, 'but it is for this reason, that it pleased men to impose both those names on the same thing.' The thesis that logical truth is due to linguistic convention and not to the existence of necessary connections in nature, has undoubtedly had very considerable influence upon the development of logic, despite the acknowledged haziness surrounding the terms 'linguistic convention' and 'necessary connection in nature'.

In the sixteenth century there are several influential textbooks worthy of notice. One of these is the *Logica Hamburgensis,* published in 1638 by Joachim Junge (1587–1657). Junge's work is mentioned by Leibniz favorably and often, usually in connection with its consideration of the so-called inferences *a recto ad obliquum.* Examples of these are the following.

> A circle is a figure; therefore, whoever draws a circle draws a figure.
> A reptile is an animal; therefore, whoever created all animals created all reptiles.

Relational inferences of this type were not Junge's discovery, however, for essentially similar cases had already been considered by Ockham; e.g.,

> All men are animals; Socrates sees a man; therefore, Socrates sees an animal.

Another such textbook was published by Arnold Geulincx in 1662, with the title 'Logic Restored to the Fundament from Which It Previously Collapsed'. It contains relatively lucid accounts of a number of standard topics, including the theory of *suppositio,* the combination of 'not' with 'all' and 'some', De Morgan's laws, and the categorical syllogism. In addition there is a discussion of the so-called antisyllogism, e.g.,

> Peter is not an animal; therefore, not both: Peter is a man and all men are animals.

The most famous and influential textbook of the period, however, was *La logique ou l'art de penser* (better known as the *Port Royal Logic*), published in 1662 by Antoine Arnauld and Pierre Nicole. With its view of logic as 'the art of correct reasoning', it is an outstanding early example of the 'how to think straight' *genre.* This is not to say that it is a bad book, but only that most of its content does not fall within the area nowadays called 'logic'. The feature for which it is best known is its distinction between the comprehension and the extension of general terms. The comprehension of a general term is the set of all those attributes that cannot be removed without destroying the concept; the extension consists of all the objects 'inferior' to the concept. In the case of the general term 'triangle', the authors tell us that the comprehension includes having extension, figure, three sides, three angles, the equality of these three angles to two right angles, etc. The extension presumably consists of all particular triangles, though the authors are not quite clear on this. It is obvious that the distinction is related to that between sense and denotation, drawn originally by the Stoics and later restated more satisfactorily by Frege.

The great philosopher and polymath Gottfried Wilhelm von Leibniz was deeply interested in logic and broached a number of ideas that anticipated developments two centuries later. His most important work remained unpublished, however, and as a result of this and other factors his influence upon the history of logic was not as great as it should have been. While still in his teens Leibniz put forward the project of constructing a *lingua philosophica* or *characteristica universalis,* an artificial language that in its structure would mirror the structure of thought. He was convinced that ordinary language, with all its ambiguity, vagueness, clumsiness, and superfluities, was not a suitable vehicle for communication or even for thinking. The idea of an artificial language was not new in itself, but Leibniz was suggesting more than just a system of notational abbreviations. The crux of his proposal was rather the notion that in the realm of thought, as in that of language, there is a complex and a simple, and that it is in principle possible to assign simple signs to the elements of thought in such manner that signs of complex thoughts are always built up in a unique way out of the signs for their parts. In such a language the linguistic expressions would be pictures, as it were, of the thoughts they represent. This, Leibniz believed, would greatly facilitate thinking and communication, and it would permit the development of mechanical rules for deciding all questions of consistency or consequence. Needless to say, Leibniz was unable to carry out his program, though he did make several attempts to construct formal calculi. In one of these he even developed part of the theory of identity, basing it on Leibniz' Law ('things are identical if they can be substituted for one another everywhere without change of truth-value'), and using the same patterns of proof we have followed in the formal system of Chapter 9, section 1. It is regrettable that he did not go further in the same direction.

The Italian mathematician Gerolamo Saccheri, known principally for his anticipation of non-Euclidean geometry, merits mention here because of his discussion of the Law of Clavius (see page 103) and his clever use of this law in proofs. His small book of logic, *Logica Demonstrativa* (1697) contains some remarkable arguments in which, to prove that a given syllogistic mood is invalid, he constructs in that mood a syllogism with true premises and a conclusion saying that the given mood is invalid. Then he argues: if the mood is valid, the conclusion of this syllogism is true and the mood is invalid; therefore, the mood is invalid. Saccheri is also noteworthy for his relatively careful use of definition; he appreciated the necessity of demonstrating the existence and uniqueness of an object before introducing a term to denote it.

More than one hundred years later we come to Bernard Bolzano (1781–1848), who wrote a huge work entitled *Wissenschaftslehre* ('Theory of Scientific Knowledge'). Scattered throughout its several volumes are various original contributions, which have only recently attracted the attention they deserve. One of these is what amounts to the first relatively sharp attempt to define analyticity and consequence in terms of interpretations. Unfortunately, Bolzano's definitions are couched in talk of propositions, as contrasted with sentences. He speaks of obtaining one proposition from another by making replacements for its constituents. But it is hardly clear what 'replacement' can mean when applied to entities purported to be non-spatial and atemporal. Consequently, though his basic idea is a good one, some substantial clarifications had to be made before it could be utilized in a rigorous way.

Bolzano defines analyticity in a wider and in a narrower sense. A proposition is universally valid with respect to a given constituent or constituents if every result of replacing these constituents by other terms is true; it is universally invalid with respect to the given constituent or constituents if every such result is false. Thus, the proposition 'The man Caius is mortal' is said to be universally valid with respect to the constituent Caius. With respect to a given constituent a proposition is analytic (in the wider sense) if it is universally valid or universally invalid; otherwise, it is synthetic. If a proposition is analytic with respect to all but its logical constituents, it is analytic in the narrower sense. This is the notion that seems to us valuable, but, because of the difficulty in making a clear distinction between constituents that are logical and those that are not, Bolzano does not place much stock in it. He goes on to define consistency essentially as follows: a group of propositions is consistent if some replacement for their non-logical constituents makes them all true. And a given proposition is a consequence of a group of propositions if it is made true by every replacement of constituents that makes true all members of the group. The close relationship between these ideas and the methods we have used in Chapter 4 is obvious.

The modern development of logic begins in earnest with the work of George Boole (1815–64) and Augustus De Morgan (1806–71). These men developed almost simultaneously the fundamentals of the so-called algebra of logic, which consists of the algebra of classes (Boolean algebra) and that of binary relations. Both Boole and De Morgan had remarked the obvious similarity of structure between certain laws of logic and corresponding formulae from ordinary numerical algebra. In working out their logical systems, however, they did not attempt to give these systems a complete characterization but only to indicate the points of difference from ordinary algebra. More satisfactory formulations of the algebra of logic were achieved only later. In the preceding chapter we have given one such formulation of the algebra of classes. A well-known and more elegant set of axioms, due in its essentials to the American mathematician E. V. Huntington, may be obtained from our set by replacing axioms 3–6 by theorems 16 and 17. The algebra of binary relations, originated by De Morgan and Charles Sanders Peirce (1839–1914), is similar to the algebra of classes except that instead of three operation symbols we have six, which are interpreted as denoting the union, intersection, complementation, relative sum, relative product, and conversion of arbitrary binary relations. Axioms for the algebra of relations have been given by Tarski.

The algebra of classes is not to be confused with the general theory of sets, of which it is only a small part. The latter, which might equally well be called the general theory of the membership relation, was created by the mathematician Georg Cantor (1845–1918). Cantor's theory incorporated analyses of the notions of cardinal and ordinal number, infinity, and many other concepts important in modern mathematics; in particular, he developed for the first time a theory of infinite cardinals, and produced a proof that for any set $\mathfrak{m}$, finite or infinite, the cardinal number of the set of its subsets is greater than the cardinal number of $\mathfrak{m}$. Cantor's theory, in its simplest form, turns out to permit the derivation of Russell's Antinomy, and consequently a very great amount of effort has been expended in attempts to find consistent subtheories that preserve as much of its content as possible.

We come now to Frege. If there is one point upon which all recent historians of logic agree, it is upon the eminent place of Gottlob Frege (1848–1925) among those who have contributed to the development of the subject. Alonzo Church says flatly that Frege 'is unquestionably the greatest logician of modern times'; I. M. Bocheński calls Frege 'undoubtedly the most distinguished thinker in the field of mathematical logic' and says that Frege's *Begriffsschrift* is comparable in importance with only one other book in the entire history of logic, namely, Aristotle's *Prior Analytics;* and William and Martha Kneale find that 'the deductive system or calculus which he elaborated is the greatest single achievement in the history of the subject.'

Frege's achievement, in a word, is that he invented logic in its modern form. In his little book, the *Begriffsschrift,* there appears for the first time a fully formalized axiomatic development of the sentential calculus, consistent and complete. Using negation and truth-functional implication as primitive, it employs six axioms, which in our notation would be theorems

4, 5, 7, 10, 11, and 15 of Chapter 6, together with *modus ponens* and substitution as the only inference rules. Even more important is Frege's introduction of quantifiers into his formal system; by including additional axioms and rules, he expands it to a complete system of first order predicate calculus. The entire presentation is in accord with his own rigorous conception of a satisfactorily formulated theory: it should be framed in an artificial, formalized language, for which the concept of (well-formed) formula is explained by reference solely to the shapes of the expressions involved; the primitive symbols must be explicitly listed, and all others defined in terms of them; all formulas to be asserted without proof are to be listed as axioms; other asserted formulas are to be derived from these by the application of formal inference rules, all of which are to be stated in advance. The number of these rules, as well as that of the axioms and primitive terms, should be kept as small as possible. What was most important of all, in Frege's view, was that the derivations be 'free of gaps'; this could be most readily achieved, he thought, if the inference rules were as few and as simple as possible.

'In making a transition [from the axioms] to a new assertion one must not be content, as up to now mathematicians have practically always been, with its appearing obviously right, but rather one must analyze it into the simple logical steps of which it consists—and often these will be more than a few'. Such remarks as this did not endear Frege to his fellow mathematicians, but he was always ready and able to give examples in which mathematical reasoning had actually gone astray because of the very deficiencies he was seeking to remedy.

Frege's other main contribution to logic goes beyond the first order predicate calculus, involving quantification upon predicate (or class) variables. This was his remarkable discovery that arithmetic, and with it other large parts of mathematics, can be reduced to logic (though not to elementary logic). The reduction is accomplished by defining the basic concepts of arithmetic in terms of purely logical notions. In its essentials Frege's method is the following. He defines any two sets as cardinally similar if there is a one-to-one correspondence between them. Then he defines the cardinal number of a set α as the set of all sets cardinally similar to α. Thus, the integer 1 is the set of all sets α satisfying the condition

$$(\exists x)(y)(y \in \alpha \leftrightarrow y = x);$$

the integer 2 is the set of all sets α satisfying the condition

$$(\exists x)(\exists y)(x \neq y \,\&\, (z)(z \in \alpha \leftrightarrow (z = x \vee z = y)));$$

and so on. The sum $p + q$ of the two integers p and q will be the set of all sets γ satisfying the condition

$$(\exists \alpha)(\exists \beta)(\alpha \in p \,\&\, \beta \in q \,\&\, \alpha \cup \beta = \gamma \,\&\, \alpha \cap \beta = \Lambda).$$

The set of positive integers may then be defined as the intersection of all sets α satisfying the condition

$$1 \in \alpha \,\&\, (n)(n \in \alpha \rightarrow \mathrm{n} + 1 \in \alpha).$$

Thus the basic concepts of arithmetic are definable in terms of logical notions only, and hence the laws of arithmetic are converted into laws of logic. In the first volume of his *Grundgesetze der Arithmetik* (1903) Frege carried through this derivation of arithmetic from logic; although his system was vulnerable to Russell's Antinomy, it can be set right in various ways, and the soundness of his basic ideas remains unchallenged.

In the philosophy of language Frege is important for his distinction of sense and denotation (see pages 22–24). His work in this area, as in every other he touched, has turned out to be extraordinarily stimulating to later investigators.

Many of Frege's insights are also to be found, less systematically worked out, in the writings of the American logician Charles Sanders Peirce (1839–1914). Entirely independently of Frege, Peirce invented a symbolism adequate for all of logic, worked out portions of the theory of quantification (even including prenex normal form), and proved important results in the theory of relations.

The next great figure in the history of logic is Bertrand Russell, who, together with Alfred North Whitehead (1861–1947), wrote the monumental work *Principia Mathematica*. Incorporating the so-called theory of types, a device designed to obviate the inconsistency of Frege's system, this three-volume treatise accomplishes in large measure Frege's program of deriving mathematics from logic. In certain respects, especially wherever the use-mention distinction is involved, it falls short of the high standard of rigor set by Frege. Nevertheless, it is without doubt a classic work and has in large part determined the subsequent development of the subject.

Finally we make brief mention of the work of Kurt Gödel and Alfred Tarski. To Gödel we owe the first proof of the completeness of elementary logic, and the still more impressive incompleteness theorem for logics of higher order. In establishing the latter theorem he showed that there can be no complete and consistent axiom system for the elementary arithmetic of natural numbers. (This theory is formulated in the first order predicate calculus with identity and operation symbols; the non-logical vocabulary consists of two binary operation symbols; the assertions are all sentences of the theory that are true when the variables are understood to range over the natural numbers and the operation symbols represent addition and multiplication.) Gödel's incompleteness theorem has had a profound effect upon the philosophy of mathematics, showing once and for all that mathematical truth cannot be identified with derivability from any particular set of axioms.

The work of Alfred Tarski has ranged over the entire field of logic, from its most philosophical to its most mathematical side. In the area of semantics, he succeeded in giving fully precise definitions for many concepts—notably that of truth—that previously had been relegated to the scrap heap of philosophical confusion; in fact, he may properly be said to have created semantics, in the scientific sense of that term. He has also made profound contributions to set theory and, in the area of metamathematics, has achieved numerous important results concerning the decidability of various mathematical theories. With his many students and associates, Tarski must be considered one of the greatest single forces moving the science of logic forward today.

EPILOGUE

Nec proinde culpandi sunt Logici quod ista sunt prosecuti, sed quod istis pueros fatigarunt.—Leibniz

BIBLIOGRAPHY

The following is a selected list of books that cover material set forth in the present text.

TEXTBOOKS

Beth, E. W. *Formal Methods*. Dordrecht, 1962.

Bocheński, I. M. *A Précis of Mathematical Logic*. Dordrecht, 1959.

Carnap, R. *Introduction to Symbolic Logic and Its Applications.* New York, 1958.

Church, A. *Introduction to Mathematical Logic,* Vol. I. Princeton, 1956.

Copi, I. *Symbolic Logic*. New York, 1954.

Curry, H. B. *Foundations of Mathematical Logic*. New York, 1963.

Fitch, F. B. *Symbolic Logic*. New York, 1952.

Hermes, H. *Einführung in die mathematische Logik*. Stuttgart, 1963.

Hilbert, D., and W. Ackermann. *Grundzüge der theoretischen Logik,* 4th edition. Berlin, 1959.

Hilbert, D., and P. Bernays. *Grundlagen der Mathematik,* 2 vols. Berlin, 1934 and 1939.

Kalish, D., and R. Montague. *Logic: Techniques of Formal Reasoning.* New York, 1964.

Kleene, S. C. *Introduction to Metamathematics*. New York, 1952.

Prior, A. N. *Formal Logic*. Oxford, 1955.

Quine, W. V. *Mathematical Logic*. New York, 1940.

Quine, W. V. *Methods of Logic*. New York, 1950.

Rescher, N. *Introduction to Logic.* New York, 1964.

Rosenbloom, P. C. *The Elements of Mathematical Logic*. New York, 1950.

Rosser, J. B. *Logic for Mathematicians*. New York, 1953.
Scholz, H., and G. Hasenjaeger. *Grundzüge der mathematischen Logik*. Berlin, 1961.
Suppes, P. *Introduction to Logic*. New York, 1957.
Tarski, A. *Introduction to Logic*. New York, 1941.

HISTORICAL WORKS

Bocheński, I. M. *A History of Formal Logic*. Notre Dame, 1961.
Boehner, P. *Medieval Logic*. Manchester, 1952.
Church, A., and others. 'Logic, History of', in the *Encyclopædia Britannica*, 1956.
Kneale, W., and M. Kneale. *The Development of Logic*. Oxford, 1962.
Łukasiewicz, J. *Aristotle's Syllogistic*. Oxford, 1951.
Moody, E. *Truth and Consequence in Medieval Logic*. Amsterdam, 1953.

INDEX